Medieval Diet and Medicine

Medical Traditions

Edited by
Alain Touwaide

Volume 10

Medieval Diet and Medicine

Occitan Health Advice for the Layperson

Edited by
Wendy Pfeffer

DE GRUYTER

ISBN 978-3-11-126376-2
e-ISBN (PDF) 978-3-11-126812-5
ISSN 2567-6938

Library of Congress Control Number: 2024933793

Bibliographic information published by the Deutsche Nationalbibliothek
The Deutsche Nationalbibliothek lists this publication in the Deutsche Nationalbibliografie; detailed bibliographic data are available on the internet at http://dnb.dnb.de.

Cover image: Collage of illustrations in medical treatises from the 1st to the 16th century CE, from Greece and Rome to the Arabic World and China (1. peaches from a Japanese edition from the Li Shi Zhen, Bencao Gangmu, A.D. 1714; 2. plant from the manuscript of Padova, Biblioteca Seminario, 194, Constantinople, ca. A.D. 1430; 3. scene of uroscopy from the manuscript of Paris, Bibliotheque nationale de France, grec, 2294, 15th century; 4. a man climbing on a tree from a manuscript of Dioscorides, Arabic, Bologna, Biblioteca universitaria, 2954, A.D. 1244).
Typesetting: Meta Systems Publishing & Printservices GmbH, Wustermark
Printing and binding: CPI books GmbH, Leck

MIX
Papier | Fördert gute Waldnutzung
FSC® C083411

www.degruyter.com

Foreword

"Eat delectably of what you want most", as "your whole life will be healed if you want to please your heart." In this sentence, which I take from the Occitan text presented here in a new, very accurate edition, one of the essential principles of medieval dietetics is concentrated: following the suggestions of the body, listening to it, supporting it in the search for pleasure. Pleasure and health, in traditional Galenic dietetics, are closely connected – therefore every dietary discussion is also a gastronomic discussion. They are connected because they are founded on a basic theorem: the flavor of things reveals their nature to us; the desire to eat expresses the needs of the body; therefore, if flavor satisfies desire – if we *like* food – this means that the nature of things (revealed by their flavor) accords with our needs (expressed by desire). All very simple, if it weren't complicated by "extraneous" suggestions, linked not to *listening* to the body but to social conventions, to cultural, ideological, psychological, extra-nutritional values that always characterize our relationship with food: it does not all take place on a material level but is definitely determined by imagination.

That warning, however, contains a notion of great interest for us too: the *responsibility* entrusted to the eaters. They must know the "scientific" rules of eating, and more generally of living, but they must also, before anything else, find *within themselves* a drive, a motivation, aware and meditated but also left to the instinct, to the impulses of the body. Not delegating everything to specialists, but *participating* in building their own diet. We will therefore regulate our diet according to the seasons, according to the passing of the months – this Occitan text also insists on the topic, a fixed point of medieval medical treatises. But we will find in the satisfaction of eating the proof that we are doing the right thing.

The basic principle of this science is always that of contrast: a "law of opposites" which has the "right middle" as its main objective. Temper the hot with the cold, the wet with the dry. Four "qualities" that arise from the combination of the four natural elements (earth, water, air, fire) characterize all living beings to varying degrees and, in turn, are "translated" into flavours. All this, albeit being universal, is absolutely individual: everyone is made in their own way, everyone needs their own food. Obviously, not everyone can afford such choices: the recipient of dietary texts (like the one you read in these pages) is always, by definition, a wealthy person. The text takes it for granted: you have a cook working for you, who, in addition to pursuing your taste, will have to take into account some general rules, such as, for example, the order of the foods: "if your cook is so crazy that he serves you the soft dish last, he should give it to you first." In short, it is a diet for the rich – as was the advice of Galen and, before him, that of Hippocrates. However, the strength of pre-modern dietetics (and the reason for its very long survival) also lay in the "banality" of its basic prescriptions: correcting hot with cold, dry with wet, are these not (also) practices suggested by daily experience? Of course, the "heat" of Galenic science is not exactly the "heat" you feel. But the idea is similar, it can be shared on a general

https://doi.org/10.1515/9783111268125-202

scale – I would even use the adjective “popular”, if I think of the many proverbs which, still in contemporary use and language, refer to the ancient science of the “humours” of the body and “quality” of foods. For this reason, the “lay person” to whom this treatise is addressed is by definition a gentleman, but by extension – and with the necessary adjustments, since people are *also socially* different – it concerns everyone. Maybe us too.

Massimo Montanari
Bologna
January 2024

Contents

Acknowledgements

The author thanks Ilaria Zamuner for giving the green light, Courtney Wells for being a first reader, Alain Touwaide for general encouragement, Nat Schwartz for multiple services, Massimo Montanari for bibliographic assistance and culinary expertise, Walter Meliga for providing materials, Marc Levinson for his photographic expertise, Kathryn Klingebiel for her red pencil, Sarah-Grace Heller for assistance with clothing technicalities, Monica H. Green for sharing her knowledge of medieval medicine, Janet Golden for pointing firmly towards medical history, and Gilda Caïti-Russo for taking needed photos. Thanks are also extended to the Interlibrary Loan librarians of Van Pelt Library, University of Pennsylvania, for helping me obtain materials not readily available, and to the Department of Francophone, Italian, and Germanic Studies (FIGS) of the University of Pennsylvania, for giving me a home.

I thank also those libraries that make available online materials of use to researchers, in particular, for this project, the Bibliothèque nationale de France, the Bibliothèque Universitaire Historique de Médecine of the Université de Montpellier, the British Library, the Real Biblioteca of the Monastery of San Lorenzo de El Escorial, the Universitätsbibliothek Basel, and the Vatican Library. I must also thank individuals who have helped me acquire the illustrations in this book, specifically Laura Boukherous, Benedicta Erny, Carlos Garcia-Minguillan, Kristina Münchow, and Jaime Sepulcre Samper.

Thanks, also, to Anne Hiller, Carla Schmidt, Lena Hummel, and others at De Gruyter, for their work in publishing this fascinating medieval text.

https://doi.org/10.1515/9783111268125-203

List of Charts and Figures

https://doi.org/10.1515/9783111268125-204

Part 1: **Presentation of *Occitan Health Advice for the Layperson***

Introduction

This volume presents a new edition of the medieval Occitan verse text dubbed the *Diététique provençale* (Provençal Dietetics) by earlier scholars[1] (incipit "Qui vol auzir un bon tractat"), and places this vernacular text in its Occitan context, that is, in the south of what is today France but what was, in the Middle Ages, an independent entity with its own language, culture, and traditions. The evocation of dietetics in the modern title brought the work to the attention of food scholars, who recognized its worth for that discipline. A fitting English title for the work would be *Occitan Health Advice for the Layperson*, the title I will use in this volume.[2]

The vernacular text is addressed to Alexander the Great and purports to offer advice first proffered by Greco-Roman physician Galen (129–after 216); as Galen lived several centuries after Alexander, the initial setting is fictional. The anonymous author tells his reader to consume certain foods only during certain times of the year, to undergo or avoid bloodletting at specific moments, and in general, he offers advice on how to lead a healthy life. *Occitan Health Advice* is as much a medical text as a source of culinary information. Some secondary works dub this text the *Epistola ad Alexandrum de dieta servanda*,[3] a title that connects the work to one of its sources but continues to conceal its medical interest. I believe the title I am proposing, *Occitan Health Advice for the Layperson* or *Occitan Health Advice* in short form, *Conselhs occitans de santat* in Occitan, offers a clearer sense of what the work tries to do. With this new title, I propose here to present this text in its fullest and to make it more available to the scholarly world.

The illustrations that accompany this edition are generally not from the manuscripts in which *Occitan Health Advice* is found. A majority of the images are illustrations in a late fourteenth-century copy of the *Tacuinum sanitatis*, now Paris, Bibliothèque nationale de France, nouvelle acquisition latine 1673.[4] Carmélia Opsomer rightly observes that the illustrations in luxury *Tacuinum* manuscripts were not inserted with documentary intent; they served medieval readers as a means of actualizing the item portrayed.[5] That these illustrations from a Latin medical work align so well with this Occitan text underscores the connection of *Occitan Health Advice* to medieval medical literature.

1 In his first presentation of the text, Suchier 1883 called the work *Provenzalische Diätetik*. Scholars have tended to use the French translation of these words, *Diététique provençale*, more often.

2 On the principle that an Occitan work should carry a title in Occitan, I elsewhere refer to this text as the *Conselhs occitans de santat*, i.e. *Occitan Health Advice*.

3 The Occitan witnesses are both called the *Epistola ad Alexandrum de dieta servanda* in the Jonas database online, see the two entries, for one prose and one for verse. While this choice of title has a logic, the Latin title for the Occitan works does not make clear their language of composition and could lead to confusion.

4 These illustrations were chosen, because they were available in open access online.

5 Opsomer 1991: 25.

https://doi.org/10.1515/9783111268125-001

Background

The history of southern France, the region where Occitan was spoken from the Middle Ages to the present day, has, for too long, been amalgamated to that of northern France. Such a mixing of regions has not served the South well. In the medieval period, the South maintained a distinctive culture along with its distinctive language. Some markers of this can be seen in the higher level of literacy in the Occitan-speaking region compared to northern France, levels of commerce closer to those of Italy than to those of less mercantile northern France, and local institutions of education unmatched in the North.[6]

Occitan Health Advice for the Layperson is part of a large group of medical texts created in the thirteenth century first in Latin and then in a vernacular language, all reflecting a growing interest in medicine in university settings and a parallel popular interest in medicine and health management.[7] Materials like *Occitan Health Advice* were "a response to a very specific social and professional demand".[8] The southern city of Montpellier's position as a center of medical education was established by the middle of the twelfth century,[9] supported by the renown of later individuals like Arnaldus de Villanova (Arnaut de Villeneuve, circa 1240–1311), who had trained there, supported as well by development of materials both for medical professionals (much of Arnaldus's oeuvre) and for a lay audience.[10]

William Rothwell suggested that the South included

> a class of medical practitioners outside the university world, not mere barbers or illiterate peddlers of old wives' concoctions, but perhaps men of the rising middle class, able to read and possibly to write their own vernacular, intelligent enough to be able to follow the indications of treatment laid down and on occasion capable of adding their own cures to the general corpus of knowledge.[11]

---

6 See Pfeffer 1997: 12–33.

7 See "Eine Diätregel für einen Bischof" (ed. Sudhoff 1923–1924) for a Latin example of these works. See Ducos 1999 for a broad survey of French and Occitan scientific translation activity. Paterson 1988: 119 offers a brief description of a circa 1218 Montpellier faculty member, Guillelmus Burgensis de Congenis (she calls him William of Congeinna; I would call him Guilhem de Congénies) and refers readers to Sudhoff 1918a: 388, and 1918b: 309, 317–318.

8 Cifuentes i Comamala 1999: 143.

9 Wallis 2010: 195. See also Paterson 1987 and her list of medical works written by authors active in Occitania prior to 1250 (Paterson 1987: 389–392). Twelfth-century French author Guiot de Provins tells his readers that those who had trained at Montpellier charge significantly for their services and lie about medications they prescribe (Guiot de Provins, *La Bible* ll. 2613–2620 [ed. Orr 1915: 91]).

10 See Cifuentes i Comamala 2004.

11 Rothwell 1983: 492–493.

In this environment, it is not surprising to find, from the thirteenth century on, vernacular works on scientific and medical topics,[12] including the text which is the focus of this book.[13]

I hasten to add that the potential sources for the Occitan text are multiple and varied; the health advice offered has its roots in a number of different traditions. As French scholar Claude de Tovar observed, the question of sources is particularly messy, involving texts that are tightly interdependent.[14]

The *Secretum secretorum* Tradition

Occitan Health Advice for the Layperson, formerly the *Diététique provençale*, has long been considered an Occitan representative of the *Secretum secretorum* tradition. The *Secretum* or *Secret of Secrets* is a text so widely copied that the historian of medieval science Lynn Thorndike (1882–1965) described it as "the most popular book in the Middle Ages".[15] The *Secretum secretorum* is itself a Latin translation of a tenth-century work written in Arabic, the *Sirr-al-asrar*, purporting to be a letter of advice written to Alexander the Great by Aristotle.[16]

---

12 Related works include the *Elucidari de las proprietatz de totas res naturals*, a translation of the encyclopedic *De rerum proprietatibus* by Bartholomaeus Anglicus (Bartholomew the Englishman), see *Elucidari*; *Lo Rosari* (an treatise on alchemy); multiple different collections of pharmaceutical recipes, the work known by its original author, the *Albucasis* or *La Chirurgie d'Albucasis*, the Occitan translation of one chapter of a Latin translation of an Arabic medical encyclopedia (the better edition is Albucasis, *La Chirurgia* [ed. Elsheikh 1992]); and *la Notomia*, a translation of part of Henri de Mondeville's Latin *Chirurgia* (Henricus de Mondeville, *La Notomia*, ed. Combes), inter alia. See Taylor 2015: 290–294. An online repertory *Repertorio informatizzato dei manoscritti di astrologia, medicina e alchimia in provenzale* (RIAMA) was announced in the early 2000's by Ilaria Zamuner; that project has since been abandoned.

13 Ducos would add that the Occitan corpus of scientific literature has a more practical, utilitarian point of view than the French corpus. "L'intérêt pour les textes à visée pratique ... Les recettes, les pronostics, les computs prouvent plutôt une lecture plus utilitaire, les connaissances encyclopédiques condensées semblent suffisantes pour comprendre le monde" (Ducos 1999: 69). Geneviève Dumas expresses surprise that "la moisson de traductions scientifiques en occitan est particulièrement pauvre à Montpellier" in Dumas 2015: 216. I think her survey of the extant literature is incomplete and that she has not taken into consideration the destruction of materials that occurred in southern France over the course of centuries.

14 Tovar 1974: 116.

15 Thorndike 1923: 2.267 attributes the line to Gaster 1907–1908, where I could not find it. I have the sense that everyone has to include this Thorndike quotation in their works; see Zamuner [Candiani] 2003: 739n2. Williams 2003a: 368–388 identified 73 manuscripts of John of Seville's translation, dating from the twelfth to the fourteenth centuries. Françoise Vielliard (2009: 228) reports 150 manuscripts. Zamuner offers a detailed history of the *Secretum* in the Romance language world in Zamuner [Candiani] 2005.

16 Gaullier-Bougassas et al. 2015a: 12; Galderisi 2011: 2.2.1251–1252.

A first, very incomplete translation of the original Arabic text into Latin prose was done by Johannes Hispaniensis (John of Seville, fl. 1109–1143) in the first half of the twelfth century.[17] Johannes dubbed his work *De Regimine sanitatis* or the *Epistola Aristotilis ad Alexandrum de dieta servanda*. This text falls in the genre of advice to princes.[18] Johannes Hispaniensis's work is known as the "short form", the shorter version, in the scholarly literature.[19] As recognized expert on the *Secretum secretorum* Steven Williams noted with regard to this effort, "the story of its *fortuna* is a complicated one" that has not yet been recounted.[20]

Translations into medieval Catalan, Castilian, Italian, and Portuguese of Johannes Hispaniensis's text are multiple; we also have two complete translations and several fragments in Old French.[21] Italian philologist Ilaria Zamuner identified Italian translations, in multiple manuscripts, not picked up by Claudio Galderisi.[22] In addition, we have this Occitan translation and its fragments.

A second and more complete translation from Arabic to Latin was done by Philippus Tripolitanus (Philip of Tripoli) in the early thirteenth century. This second translation, known as the "long form", the *Secretum secretorum*, was widely copied and translated into multiple vernaculars over the next few centuries, though not into Occitan.[23] Philippus Tripolitanus's translation, a longer text in the genre of medicine and/or of magic, has received much more scholarly attention than the shorter version.

The *Secret of Secrets*, in its fullest, i.e., the version by Philippus Tripolitanus, consists of ten books[24]:

Preface
1. On the Kinds of Kings
2. On the Position and Character of a King
3. On Justice
4. On Ministers
5. On Scribes
6. On Ambassadors

---

17 Zamuner [Candiani] 2003: 742–743; Bizzari 2014: 44.

18 This text was described in Suchier 1883: 531. Marilyn Nicoud calls it a treatise on hygiene rather than a medical text (Nicoud 2007: 958) and does not really include *Occitan Health Advice* in her study.

19 See inter alia, Galderisi 2011: 2.2.1252.

20 Williams 2003b: 469.

21 See Monfrin 1982, and Zamuner [Candiani] 2005 and 2006 for a detailed discussion of this tradition. Tony Hunt's concern that this translation tradition had not yet been adequately examined, expressed in Hunt 1999: 351, is perhaps now somewhat assuaged. Distinguishing translations of Johannes Hispaniensis's text from those of Philippus Tripolitanus's is not as simple as might be wished, no surprise for this massively copied Latin work.

22 Zamuner [Candiani] 2003: 744 and 2021; Galderisi 2011: 2.2.1253–1254.

23 See Williams 2003a and 2003b: 469–482; Vielliard 2009: 228; Galderisi 2011: 2.2.1252.

24 Williams 2003a: 10–11.

7. On Governors
8. On Army Officers
9. On the Conduct of War
10. On the Occult Sciences

Of these, Johannes Hispaniensis's translation includes only elements of Book 2's "Section on Health", whose long version includes[25]:

Introduction to the Section on Health
Practical Recommendations on Hygiene, Food/Drink, Exercise/Rest; the Four Seasons
The Four Parts of the Body
Diet (Preserving Natural Heat; Healthy Foods; Meat and Fish)
Water
Wine
The Bath
Panacea
Medical Astrology

Johannes Hispaniensis reduces these nine headers significantly,[26] presenting:

Introduction with an Explanation of the Translation Sequence (Greek to Arabic to Latin)
Recommendations on Hygiene, Food and Drink, Exercise and Rest
The Four Seasons
Preserving Natural Heat

Occitan Health Advice includes not only these elements, but incorporates significant information from other sources, most clearly the diet calendar tradition (see below), and concludes with a theme found in troubadour literature, the importance of joy and pleasure.

Galenic Tradition and Humoral Theory

Underlying *Occitan Health Advice for the Layperson* is the foundational theory of medieval medicine, the belief that four "humors" govern illness and health. This theory of humors finds its first enunciation by Greek physician Hippocrates (460 B.C.–between

---

25 Williams 2003a: 10.
26 As published by Brinckmann, they make up 213 lines in total in Brinckmann 1914: 39–46; as published by Suchier, they make up 231 lines in total, see Suchier 1883: 473–480.

375 and 350 B.C.) in a work associated with him, *On the Nature of Man*.[27] The theory was formalized by the Greek physician, natural scientist, and philosopher Galen, whose career took him from his birthplace in Pergamon (Asia Minor, today's Turkey), to Rome, where he settled, wrote, and practiced medicine.

> Galen ... conceives of things as composed of the four elements of fire, air, earth, and water, formed by the union of matter, and the four qualities of hot, cold, dry, and moist. ... The food and drink animals consume consist of these elements. In the process of digestion, food and drink turn into the bodily juices, the humors, of which there are four main kinds: blood, phlegm, yellow bile and black bile.[28]

These humors lend their names to four temperaments: sanguine, choleric, melancholic, and phlegmatic. An imbalance of humors was thought to be the cause of all diseases; health was the result of humoral equilibrium. Galen was a "dietetic physician" who considered that "Most internal diseases [were] caused by errors of regimen, and hence avoidable".[29]

The qualities of warm, cold, moist, and dry, and the four humors were connected to the four seasons: blood (hot and wet) to spring, yellow bile (hot and dry) to summer, black bile (cold and dry) to autumn, and phlegm (cold and wet) to winter, creating an arrangement well laid out by food historian Melitta Weiss Adamson[30]:

Chart 1: Qualities, Humors, and Seasons (based on Adamson 1995: 11).

	blood hot and wet spring	
yellow bile hot and dry summer		phlegm cold and wet winter night
	black bile cold and dry autumn	

Galen combined elements of medical theory from a number of sources and added levels or degrees of intensity[31] that vary from mild to strong on a scale of one to four

27 Adamson 1995: 11. See Temkin 1973: 103, "The emphasis on these four humors as *the* Hippocratic humors, the linking of them with the Aristotelian qualities and with the tissues of the body was largely Galenic" (Temkin's italics).

28 Temkin 1973: 17.

29 Temkin 1973: 40.

30 Adamson 1995: 11.

31 See Temkin 1973: 15–20; Adamson 1995: 16; Green 2020: 378. The example of basil is taken from Ibn Butlan, *Taqwim al-Sihha* (ed. Elkhadem 1990: 164).

(basil, for example, could be described as warm to the second degree and dry to the first degree).[32] His principles would be followed well into the modern era, and his reputation as a physician held strong for years. "His name had a grip on the popular imagination."[33] Galen's reputational standing explains the premise of this *Occitan Health Advice for the Layperson*, that it is a text written by the Greek physician (who lived between 129 A.D. until possibly after 216 A.D.) to a Greek emperor, Alexander the Great, who lived more than 500 years before him (356–323 B.C.) (see ll. 25–28).[34]

The Baghdadi author Ibn Butlan (died circa 1065) further codified the principles of Galenic theory, presenting these in tabular form and thereby creating a practical handbook, the *Taqwīm as-Siḥḥa*. Ibn Butlan expanded the concept of levels of intensity to include much beyond the vegetable world, such as meat and animal products, the seasons, winds, aromatic substances, and habitations.[35] His work in Arabic was translated into Latin in Palermo during the second half of the thirteenth century, taking the title of *Tacuinum sanitatis*.[36] This work became rapidly popular in the late Middle Ages, and multiple copies are extant. Illustrations from some of the finer copies are often used to illustrate medieval food history discussions.[37] Because of the tabular nature of this work, it is an easy reference point for information on specific items included. Even in the versions of this work that are more illustration than text, we find, for each item, a summary description of its nature and degree, what is considered the best variety, its utility, harmfulness, and the way to minimize the harm (below, the entry for limes or lemons, from a copy of the *Tacuinum sanitatis* now in Paris).

A different text expressing similar ideas is the *Regimen sanitatis Salernitanum*, twelfth- or thirteenth-century health advice associated with the School of Salerno, a renowned medical school in the Middle Ages,[38] though the work may not have originated there.[39] Composed in Latin leonine hexameters, initially 364 lines long, though

32 Temkin 1973: 112 offers a useful description of these differences in intensity as they relate to drugs: "the first where the action was inferred theoretically only, the second where it was perceptible, the third where the effect was strong, and the fourth where it was destructive".

33 Temkin 1973: 125. Galen's name appears in Shakespeare's *Coriolanus*, Act II, scene 1, for example.

34 Kibre 1945: 403–406 discusses these letters briefly, connecting the Hippocratic *Letter to Antiochus* to the Aristotelian *Letter to Alexander the Great*.

35 Adamson 1995: 91.

36 Adamson 1995: 83.

37 Adamson reports seventeen extant medieval Latin copies, Adamson 1995: 84. It is most often an abbreviated version that is lavishly illustrated, see, for example the manuscripts in Vienna, Osterreichische Nationalbibliothek, *Codex Vindobonensis series nova* 2644, in Rome, Biblioteca Casanatense, codex 4182, and in Liège, Bibliothèque universitaire, manuscript 1041.

38 "By the turn of the twelfth century, 'Salerno' became almost a brand name for learned (and often expensive) medicine practiced by scholarly physicians trained in the physiology and therapeutics of Hippocrates, Galen, and Avicenna," Black 2020: 184. See also Van d'Elden 1991: 503–505.

39 Black 2020: 185.

Fig. 1: Limes or Lemons, *Tacuinum sanitatis*, Paris, Bibliothèque nationale de France, nouvelle acquisition latine 1673, f. 18r. Source: gallica.bnf.fr.

the work grew to 3,526 verses over time,[40] it is sometimes and wrongly attributed to Arnaldus de Villanova, who did write a prose commentary on the text. This work was immensely popular in the Middle Ages and afterward, with translations into a host of medieval and modern languages.[41] The *Regimen* has been described as "a highly compressed health-book" with close affinities to the *Secretum secretorum*: general considerations of health, the morning routine, general aspects of nutrition and specific foods, along with discussion of the four seasons and of blood-letting.[42] The *Regimen* makes clear its stand on humoral theory:

> Quatuor humores in humano corpore constant:
> Sanguis cum cholera, phlegma, melancholia.
> Terra melancholia, aqua phlegma, aer sanguis, cholera ignis.
>
> (Four humors make up the human body:
> Blood, choler, phlegm and melancholy.
> Earth [corresponds] to melancholy, water to phlegm, air to blood, fire to choler).[43]

The *Regimen sanitatis* added another approach to Galenic theory, referring to "non-naturals", six factors that influence health and illness: air, physical exercise, diet, sound, bodily excretions, and emotions. Some of these factors are explored in the Occitan text.

We see evidence of familiarity with Galenic theory in many medieval Occitan texts. For example, a thirteenth-century Montpellier pharmaceutical recipe collection enunciates the four humors clearly:

> catre humors: sanc, colera, fflecma, malencolia. Lo sanc es semblans ad aire, cautz et humoros; la colera, a fuec, cauda e seca, fflecma, a l'aiga, freja et humoroza; malenconia, a la terra, freja et secca.

---

40 Adamson 1995: 97.

41 Adamson includes a "Provençal" printed edition among the numerous translations and printings (Adamson 1995: 98), citing Fielding Garrison, who mentions this translation without providing any documentation whatsoever, Garrison 1929: 150.

42 Adamson 1995: 101, offers a chart presenting the correspondences between the *Regimen Salernitanum* and the *Secretum secretorum*:

Regimen Salernitanum		*Secretum*
[Cap. 2]	De confortatione cerebri	[Cap. 30]
[Cap. 3]	De somno meridian	[Cap. 32]
[Cap. 6]	De dispositione ante cibi sumptionem	[Cap. 35]
[Cap. 7]	De cibis melancholicis vitandis	[Cap. 50]
[Cap. 30]	De piscibus	[Cap. 50]
[Cap. 80]	De nocumentis visus	[Cap. 46]
[Cap. 87]	De quatuor temporibus anni	[Cap. 37–40]

43 *Regola sanitaria salernitana* (Engl. transl. Gherli 198–: 72; reprinted in Matterer 2001: 13). See also Black 2020: 193, and Cummins 1976.

(four humors: blood, choler, phlegm, melancholy. Blood is similar to air, warm and moist; choler to fire, hot and dry; phlegm to water, cold and moist; melancholy to earth, cold and dry).[44]

Matfre Ermengaud († 1322) expresses the concept of degree clearly when he states:

Atressi cove que veja
l'erba si es cauda o freja,
et en qual gra determenat
es en la dicha qualitat,
premier, segon o ters o quart.

(So it is appropriate to consider
the herb, if it is hot or cold,
and to which degree
in this quality:
first, second, third or fourth).[45]

These Galenic principles were generally accepted in the medieval period and by the author of *Occitan Health Advice for the Layperson.*

Diet Calendars

There exists another genre of Latin advice literature relevant to this Occitan text, works which I will call "diet calendars", relatively short prose texts organized by the months of the year offering health advice; they may not yet have been examined in full,[46] though the texts are numerous in medieval manuscripts, particularly of the eleventh and twelfth centuries.[47] These calendars carry various titles, including *Regimina XII mensium* and *Medicina Ypocratis quid usitare debeatur per singulos menses.*[48] I offer here some information about these works, because they are a source for *Occitan Health Advice.*

---

44 Brunel 1957: 294–295. Here and throughout, all translations are by the author unless otherwise indicated.

45 Matfre Ermengaud, *Breviari 2* (ed. Ricketts 1989: lines 7119–7124).

46 Fischer 2015: 112n3, names the many medical historians who have ignored these works; he also points to Sabbah et al. 1987, no. 82, and cites Groenke 1986. In contrast, Wallis states, "Medical calendars of this type are very common in medieval manuscripts and have been studied extensively", Wallis 1995: 113. It is surprising that Fischer was unaware of Pucci Donati 2007's careful study of this material.

47 Groenke identifies 70 different Latin witnesses which he divides into 9 types in Groenke 1986. I use Groenke's types from this point forward. Note that Ortrun Riha 1987: 377, suggests there are not nine but twelve different types.

48 Fischer 2000: 10; Zamuner [Candiani] 2006: 172. *Medicina Ypocratis quid usitare debeatur per singulos menses* is the title used by the scribe of British Library Harley manuscript 3271, f. 122 verso. The Bibliothèque nationale de France uses the title "Liber de minucione et medicina secundum variationes lunarum", found in thirteenth-century manuscript Bibliothèque nationale de France, fonds latin 10448, f. 122 recto (see Gallica online) and published in Sudhoff 1908. Beccaria uses "calendario dieteti-

French pharmacist and historian of pharmacy Jean Barbaud (1918–2008) proposed that the source of these works was Pseudo-Soranus, a turn-of-the-sixth-century text, an idea rejected by medical historian Faith Wallis (1950–).[49] A different individual, Soranus of Ephesus (2nd century A.D.), would find his name associated with a work published in the Renaissance with the title *Isagoge in artem medendi*,[50] an *Introduction to the Art of Healing*. Klaus-Dietrich Fischer argues that the elements of the *Isagoge* printed in 1528 may have circulated independently, many of them anonymously, prior to their being assembled into a single volume whose works would be attributed to a single author, Soranus.[51] Dutch historian of medicine Johanna Van Winter (1927–) argues that the "tradition of pseudo-Soranos ... does not play a role in the diffusion of the health rules in popular medicine".[52] She traces the origin of these texts to fourth-century physician Theodorus Priscianus whose work was combined with elements of the *Secretum secretorum*, which included a section on diet according to the seasons.[53]

We need to add to the list of potential sources for these calendars examples of Late-Classical pseudo-epigraphic literature, specifically letters attributed to renowned Greek physician Hippocrates of Kos (circa 460 B.C.–circa 370 B.C.) to various recipients, including Antiochus (fl. 4th century B.C.),[54] in which we find something of a health calendar. The letter to Antiochus was subsequently freely adapted into a letter to Maecenas (circa 70–8 B.C.), the *Epistola Hippocratis ad Maecenatem*.[55] Based on dates alone, it is clear that the letter to Maecenas must be a pseudo-Hippocratic text.[56] Paleographer Carmélia Opsomer suggests that the Latin title of *Epistola Hippocratis*

co" to describe these texts; he identified some fifteen witnesses, with varying incipits, in manuscripts from the ninth through the eleventh century, a number which suggests that the work was well known in that period, see Beccaria 1956: 443. Thorndike and Kibre 1963 add to Beccaria's list. See Barbaud and Gillon 1988 for a discussion of these then relatively unstudied texts. Groenke 1986 is the first attempt to present a critical edition of the various versions of these monthly calendars; he counts seventy of them, divided into nine distinct types (Groenke missed at least one; see Stuart 1979, mentioned in Fischer 2000a: 18). Note also Zamuner [Candiani] 2006: 172. For Groenke Type IIa, the German scholar counted no fewer than twenty-six examples, Fischer 2015: 117, citing Groenke 1986: 45. Ortrun Riha enumerated some of the multiple German-language examples in Riha 1994.

49 Barbaud 1988: 332, and 1986: 21; Wallis 1995: 135n22.

50 See Fischer 2000b.

51 Fischer 2000b: 7–8 and 20. See also Green 2019 for a similar assembly process used in Renaissance publication of medieval gynecological texts.

52 Van Winter 2010: 99.

53 Van Winter 2010: 108.

54 See Opsomer and Halleux 1985 for an overview of these influential letters. Opsomer and Halleux 1985: 349 describe this work as "un texte très mouvant". They offer several other titles, including *Epistola Hippocratis ad Antiochum, Epistola Ypocratis medici ad Antiochum* ..., and *De morbis quattuor regionum corporis*.

55 Opsomer and Halleux 1985: 349, note that other titles exist, including *Liber medicinalis Ypocratis, Epistola medicinalis quatuor temporum* and *Liber dinamidiorum, Galenus ad Maecenatem de regenda santitate*.

56 Talbot 1972: 27, offers an origin narrative for this text.

translata ad Maecenatem should be understood as "Hippocrates' letter translated for Maecenas".[57] Pearl Kibre (1900–1985) describes these epistles as "providing in convenient form ... dietary and medical or hygienic rules of health".[58]

Chart 2: Comparison of *Occitan Health Advice* and Latin *Letter of Hippocrates*.

Occitan Health Advice	*Antiocho regi Hippocrates Cous salutem*[59]
Apres ven hyvern ab lo freig que moutas res ten en destreig. Adonchas deu hom pron manjar, e·l cors moure e escalfar, et esser pres de la cosina; e manjar tota salvasina; e far raustir sobre·ls carbons gallinas, e gails, e capons, e manjaras raust e panadas, aste de porc, e carbonadas que sian trastug salpicat d'especias e empebrat; e beu bons vins et bons pigmens per contrastar als elemens. E rescon sotz ton cobertor bela domna ab fresca color – e non poinhes d'autra mesina adoncs; que non i a tan fina.	Ex die VIII Kal. Januar. corporibus umor adcrescit usque veris aequinoctium, quod incipit VIII Kal. April. Hoc ergo tempore utendum calidissimis cibis et optimis, vino aliquatenus indulgendum, usus etiam venerius minime respuendus.
Afterwards comes winter with the cold which keeps many people in distress. Then must one eat a lot, and move and heat the body, and stay close to the kitchen; and eat all kinds of game; roast over coals chicken and quail and capon, and you will eat roasts and empanadas, skewers of pork and grilled meats that are completely sprinkled with spices and with pepper; and drink good wines and good piments to contrast with the elements. Hide under your bedcovers a lovely lady with a good complexion – take no other medicine at that time, for there is none finer.	From the 8th day of the calends of January moisture accumulates in bodies until the vernal equinoxes, which begins the 8th day of the calends of April. At this time, then, the hottest and best foods should be used, wine should be indulged to some extent, and sexual practices should not be rejected in the least.

57 Opsomer 1991: 15. See also Sabbah et al. 1987: 99–100.

58 Kibre 1945: 404.

59 Text in Helmreich 1889: 8.

Certainly, the Antiochus letter offers a seasonal calendar, starting with winter and including advice on activities and foods according to the season. It is hard to correlate the information in this Latin text directly with that provided by *Occitan Health Advice* because the differences far exceed the similarities: for one, the Occitan text starts the year with spring. The chart above compares only the winter season, from which it is very clear how the Occitan author expanded on this potential Latin source.

The advice on eating warm foods, drinking wine, and enjoying sex is maintained, but much expanded in the vernacular presentation. While not visible in this example, the Latin text's use of the Roman calendar has its echoes in our Occitan text as well. I note that a version of this Latin work is found in University of Montpellier manuscript Bibliothèque Universitaire Historique de Médecine, H 185[60] (see above).

In addition, a multitude of vernacular medical texts carried the idea of a letter from Hippocrates forward; many of these are called the *Letter from Hippocrates to Caesar*, accurately described by British philologist Tony Hunt as "a somewhat amorphous treatise ... which has attracted the name 'Lettre d'Hippocrate' as a critical convenience.[61] These largely vernacular works[62] (some of which have little one-to-one correspondence to the Latin *Epistola*) could include information on the four humors and on interpretation of urine, along with pharmaceutical recipes for a variety of illnesses. There may be six extant Occitan examples,[63] but these present only recipes of treatments, without any other medical discussion.[64] Some French versions do offer advice on diet that follows the calendar;[65] the connection between these vernacular texts and the Latin *Regimina XII mensium* remains to be studied.

Despite the uncertainties that remain as to the origins of medieval diet calendars, what is clear today is that there were, at a minimum, two regimens tied to the calendar, one according to the seasons and a second according to the months of the year.

60 *Regimen XII mensium*: Montpellier, Bibliothèque Universitaire Historique de Médecine, Université de Montpellier, H 185, ff. 160 recto–160 verso.

61 Hunt 2011: 17, suggests that works under the title of "Lettre d'Hippocrate" originate in vernacular languages. The scholar adds, "no Latin source has been recorded, ... the collection as a whole has seemed to be an essentially vernacular work", Hunt 1990: 100–101. He argues that the Latin "Lettre" included in London, British Library, Royal 12 B XII is a translation of a French version of the work (Hunt 1990: 101).

62 Hunt reports one Latin witness, London, British Library Royal 12 B XII, ff. 171 verso–175 verso, published in Hunt 1990: 124–136 and 365 (see note 61). There are close to fifty witnesses in Old French and in Anglo-Norman, see Archives de littérature du Moyen Âge (known as ARLIMA), s. v. *Lettre d'Hippocrate à César*. For the multiple copies in Middle English, see Archives de littérature du Moyen Âge, s. v. *Letter of Ipocras*.

63 See *Dictionary of Old Occitan Medico-Botanical Terminology* online.

64 The text is found in Auch, Archives départementales du Gers, manuscrit I 4066, ff. 67–79. See Brunel 1956 and 1935: 21. Francisque Michel (1809–1887), a French historian and philologist, owned a manuscript which included an Occitan *Letter of Hippocrates*; his copy appears to have been lost (see Brunel 1956: 81n2).

65 See Hunt 2011: 100–102.

German medical historian Frank-Dieter Groenke (1958–) distinguishes nine different variations of the monthly calendars in his published 1986 dissertation.[66] Van Winter observes,

> [I]n the month-based regimens the medical aspect dominated and in the seasonal regimens it was the culinary one which was emphasized; the medical aspect of the month-based rules was especially directed towards bodily purging.[67]

Whether by month or by season, these regimens outline good health habits for the entire year; some monthly versions start the year in January, others in March.[68] Van Winter distinguishes between those that advise absinthe in May as opposed to those that recommend absinthe for April.[69] Both variations are frequently inserted into other medical works. The scholarship on these works suggests a diffusion out of Insular monasteries in the eighth century (Benedictus or Benedict Biscop [circa 628–690] and Beda Venerabilis or Bede the Venerable [672–735] are implicated in the transmission of these regimens)[70] though Van Winter suggests Salerno as another source;[71] Groenke points to Ravenna.[72]

These diet calendars succeed in concealing themselves from the modern reader and cataloguer;[73] they are hiding in medical texts in libraries in Europe and North America today.[74] Important for *Occitan Health Advice for the Layperson* is that two monthly calendars are found in a codex (Montpellier, Bibliothèque Universitaire His-

---

66 Groenke 1986, but see Riha 1987: 377. One could argue that Groenke's Type VI, for which he presents two witnesses (Groenke 1986: 219–223), both from the same manuscript, Vatican Library, *Reginenses latini*, 1143, is more a seasonal regimen than a monthly guide.

67 Van Winter 2010: 108.

68 University of Montpellier manuscript H 185 offers an example of each of these types, Groenke Type I, which begins with January, on f. 160 recto–verso; his Type III, which begins with March, on f. 161 recto–verso; see Groenke 1986: 97 and 167.

69 Van Winter 2010: 101. Groenke Type I and Type III offer examples of this difference in prescription.

70 Barbaud 1986: 21. Beda is invoked in Groenke Type V, incipit "Observatio mensium secundum Bedam", Groenke 1986: 210. Braekman points to Beda's *Computus vulgaris qui dicitur Ephereris* (Braekman 1986: 56–57).

71 Van Winter 2010: 108.

72 Groenke 1986: 278.

73 The online catalogue descriptions of University of Montpellier, Bibliothèque Universitaire Historique de Médecine, H 185 offer incomplete lists of contents, in which this text does not appear. See the online catalogue of French university libraries, Catalogue en ligne des archives et des manuscrits de l'enseignement supérieur (Calames), which picks up information from the library's own online catalogue, the Réseau des bibliothèques universitaires de Montpellier. The calendars are hiding under item 16, *Antidotarium*, the header used by paleographer Miriam Drabkin, who published a description of the script and partial transcription of this text in Drabkin 1942: 425–426. The Jonas database online offers a minimal description and no contents for this manuscript. Groenke catalogues one of these as his type I (Groenke 1986: 21), the other as his type III (Groenke 1986: 167). See also note 68 above.

74 See Groenke 1986: 21, 45, 76, 80 and 91–95.

torique de Médecine, Université de Montpellier, H 185)[75] that was held in a monastery near Avignon in the fourteenth century,[76] confirming that this month-by-month method of presentation of medical information was available in Occitania at the approximate time of composition of *Occitan Health Advice.*

I hypothesize that the anonymous author of *Occitan Health Advice* consulted the manuscript now in Montpellier; certainly his work adopts some of the advice provided in the calendars found in that codex (see below). While the evidence is circumstantial, it does suggest that the Occitan author knew this Latin work or another much like it.

The Montpellier School of Medicine and its Faculty

It is useful to add information here on formal medical education in medieval Occitania, to provide additional context for vernacular developments in medical information.[77] The twelfth century knew of two important medical schools, one in Salerno, discussed above, and a second in Montpellier. The faculty in Montpellier contributed significantly to the prestige of the school; of these, the most renowned was a Catalan native, Arnaldus de Villanova.

Arnaldus de Villanova or Arnaut de Villeneuve (circa 1240–1311) is one of the most prominent figures in medieval medicine. Originally from Catalonia, he trained in Montpellier and was on the medical faculty there from 1289–1299. He also traveled to the courts of Aragon, Sicily, France, and Naples, to the cities of Rome and to Avignon. His medical works are dedicated to several noble patrons and patients, including Robert of Anjou, king of Naples; for a while he was protected by Pope Clement V. Important for *Occitan Health Advice for the Layperson* is his *Regimen sanitatis ad regem Aragonum*, written circa 1308 in Montpellier[78] and quickly translated into Catalan.[79] Arnaldus used the non-naturals (see above) as the organizing principle of his

75 Groenke 1986 distinguishes these calendars as examples of his Type I (f. 160 recto–verso) and Type III (f. 161 recto–verso).

76 Though the manuscript dates from the eleventh century, f. 1 recto bears the phrase "Iste liber est monasterii sancti Andree dyocesis Avinionensis" in a fourteenth-century hand; see Pansier 1907: 116 for a discussion of how the manuscript may have arrived in Montpellier; the online Catalogue en ligne des archives et des manuscrits de l'enseignement supérieur offers no more than the provenance of a Benedictine abbey in Saint-André de Villeneuve-lès-Avignon.

77 I regret that I learned of *Gui de Chauliac et sa Grande Chirurgie*, edited by Béatrice Bakhouche, Evelyne Berriot-Salvadore and Daniel Le Blévec (2023), too late to incorporate findings from that collection of essays to this volume.

78 Adamson 1995: 110. Arnaldus de Villanova, *Regimen sanitatis ad regem Aragonum* (ed. García-Ballester and McVaugh 1996).

79 See Gil-Sotres et al. 1996: 409–416, for the translation history.

treatise,[80] directed to a non-specialist audience: King James II of Aragon (1267–1327) and his court. Medieval historian Lluís Cifuentes has noted that copies of Catalan translations of Arnaldus de Villanova's *Regimen sanitatis* were owned by commoners, noblemen, members of the clergy, along with members of the medical profession (doctors, surgeons, and apothecaries),[81] an indication of how medical information moved quickly out of a learned setting both in Catalonia[82] and in Occitania.

Arnaldus was hardly the only medical professional associated with Montpellier; his near-contemporary Bernardus Gordonius (Bernard de Gordon, fl. 1283–1320) published his *Regimen sanitatis*, part of his *Liber de conservacione vite humane*, in 1308, a work of four books using Galenic theory and covering many aspects of health maintenance; this work was directed to a specialized audience.[83] Towards the middle of the fourteenth century, Stephanus Aldebaldi (fl. 1340–1350), another Montpellier faculty member, was also publishing medical texts for non-specialist readers, notably a surgical text.[84]

The Spread of Knowledge from Latin to the Vernacular

Maria Sofia Corradini Bozzi, an Italian philologist with a specialization in pharmaceutical vocabulary, considers how medical information passed throughout the medieval Mediterranean world in her essay "La letteratura medica medievale in lingua d'oc", seconding ideas presented by medical historian Danielle Jacquart[85] and pointing to Montpellier and Toulouse as two important centers of medical education.[86]

Corradini Bozzi suggests that Montpellier continued the Salernitan model. However, Michael McVaugh describes a shift in the Montpellier curriculum in the first third of the fourteenth century;[87] Danielle Jacquart argues that there was a shift in the underlying thinking of teachers, aligning Galenic and Aristotelian philosophy in this period.[88] Melitta Weiss Adamson argues that with the growing importance of Montpellier as a medical center, the "gradus system",[89] the ranking of food items in

80 Carré 2019: 175.
81 Cifuentes i Comamala 1999: 140, Carré 2019: 175.
82 See Gil-Sotres et al. 1996: 409–416.
83 Gil-Sotres et al. 1996: 81. See Demaitre 1980: 59–70 and 2004.
84 Rinoldi 2021: 190, suggests that Aldebaldi wrote his text in Occitan.
85 See Jacquart 1998.
86 Corradini Bozzi 2012: 151.
87 McVaugh 2017.
88 Jacquart 1998: 217–220.
89 Bernardus Gordonius's *Tractatus de gradibus*, completed in 1303, may be a final example of the earlier theory. See Demaitre 1980: 48–51.

terms of degree, was slowly abandoned.[90] She observes a movement towards a "regula system", based on rules for consumption of foods, suggesting that this rule-based system was used by faculty and students in medical schools in Montpellier and Paris as a pedagogical tool to help master the material.[91] Marilyn Nicoud describes the first half of the fourteenth century as the Golden Age of the Montpellier school of medicine, a period when we see "a fairly great diversity of writings, dictated by a concern to propose regimens for a variety of readers".[92] Nicoud points to the works of Arnaldus de Villanova, Jordanus de Turre (fl. 1313–1335), Gerardus de Solo (also known as Gerardus Bituricensis,[93] at Montpellier between 1335–1337), Raymondus de Molière (fl. 1335), Gerardus de Marcerii (at Montpellier between 1337–1342) and Bernardus Gordonius (Bernard de Gordon) before discussing the Latin texts that offer dietary advice for good health composed by Stephanus Arlandi (fl. 1319), Johannes de Bononia (Jean de Bologne) and Arnaldus de Bamberga or Arnold of Bamberg (fl. 1314–1321),[94] not all of whom were directly associated with Montpellier. Nicoud argues that dietary advice was an important part of the Montpellier curriculum and that authors found a ready audience, a readership interested in medical topics.[95]

Paolo Rinoldi speaks of "manuali 'migranti'", wandering manuals, that carried Montpellier's teaching across Europe.[96] Certainly the Montpellier medical school was recognized as an authority and worthy of invocation.[97] Contemporary with this shift is the spread of the *Secretum secretorum* and the *Regimen sanitatis Salernitatum* texts.[98] These works reflect concepts circulating throughout the medieval world, erasing the line between medical and dietary advice and crossing into the domain of the cookbook.[99] Claude de Tovar agrees with the idea of fluidity of genres and argues that as early as the thirteenth century, we find an effort to provide texts in vernacular

---

90 Adamson 1995: 200.

91 Adamson 1995: 201. The early fourteenth-century Latin text attributed to four Montpellier faculty members does not seem to use either Galenic or regula-based theory explicitly for its counsel. See "Eine Diätregel für einen Bishof" (ed. Sudhoff 1923–1924).

92 "Une assez grande diversité d'écriture, dictée souvent par le souci de proposer à un public varié des régimes adaptés à son mode de vie", Nicoud 2004: 234.

93 See Guénoun 1999: 465–466.

94 See Kallinich and Figala 1972.

95 "Un public intéressé par la 'chose médicale'", Nicoud 2004: 240. Zamuner also argues that *Occitan Health Advice* was intended for a university audience, "s'insérant finalement assez maturellement dans l'*Articella*, le corpus destiné à la formation des étudiants de médecine," Zamuner [Candiani] 2006: 173. Cifuentes identifies at least two distinct audiences for medical recipe collections, one professional but outside the university setting, the other "domestic", readers with a personal or domestic use independent of the professional medical community; see Cifuentes i Comamala 2016: especially 132–145.

96 Rinoldi 2021: 179. See also McVaugh 1976.

97 Braekman 1986: 67, calls attention to the incipit of one dietary, "This was writen fro Mou*n*pileris to Quene Isabel of ynglond, at þe preier of ye kyng of Fraunce, hir broþer".

98 Adamson 1995: 201.

99 See Carré 2019: 174; Gil-Sotres et al. 1996: 202–205.

languages to a broad public, literate and illiterate, certainly ignorant of Latin. She points to a common character for all these texts, "a concern for popularization and for practical use".[100] Robert Halleux and Carmélia Opsomer observe that the fact of translation from Latin into a vernacular language underlines not only a change in audience but also a change in format, so as to provide practical, efficacious, and immediately useable material,[101] what François Bérier calls "utilitarian works".[102] Significantly, it appears that translations of medical works may have appeared in Occitan before they were created in other vernacular languages.[103]

The second half of the fourteenth century saw a significant increase in vernacular texts offering health and dietary advice;[104] these texts share several features which differentiate them from their predecessors. The first difference is that the more recent works show a tendency towards brevity – they are comparatively short (two or three manuscript pages). Their focus on the non-naturals as an organizing principle[105] distinguishes these texts from the *Occitan Health Advice*, probably composed somewhat earlier and organized more on the calendar. Another feature of the late medieval corpus of health literature is anonymity. It is postulated that "anonymity is a logical consequence of the process of generalization of medieval hygienic literature".[106] Jennifer Borland postulates a network of female readers for the regimen texts,[107] but nothing in our text nor its several witnesses points in that direction.

The regimen literature is useful for food historians at several levels, providing information ranging from "general aspects of nutrition, the dietetic qualities of foodstuffs, to actual culinary recipes".[108] While *Occitan Health Advice for the Layperson* certainly fits into this genre, it does not provide recipes, placing the text closer to the *Secretorum* tradition than to the regimen tradition. Nonetheless, the work incorporates elements of both.

100 Tovar 1974: 115, "le souci de vulgarisation et d'utilisation pratique".

101 Halleux and Opsomer 2005: 344.

102 Bérier 1988: 234. Hunt 1994 speaks of "the flourishing activity of medical writings in southern France before 1250" and points the reader to Paterson 1987.

103 See Segre 1968: 143. Cambridge, Trinity College, manuscript 0.1.20, a collection of vernacular medical works, is dated by Hunt to the period 1230–1260 (Hunt 1994: 1.17), suggesting French or Anglo-Norman interest in medical material as early as the mid thirteenth-century.

104 See Gil-Sotres et al. 1996: 101–105, for some of these works.

105 Gil-Sotres et al. 1996: 89.

106 Gil-Sotres et al. 1996: 95, "la freqüència de l'anonimat ha d'interpretar-se com una conseqüència lògica del procés de generalització de la literatura higiènica medieval."

107 Borland 2022: 91, among others.

108 Adamson 1995: 201.

Direct Sources for *Occitan Health Advice for the Layperson*

At some point in the thirteenth century, an anonymous Occitan author used Johannes Hispaniensis's work as the rough basis for a work of his own, which begins “Qui vol auzir un bon tractat”. I do not think we can call the Occitan advice a translation of Johannes Hispaniensis, though large sections of the Latin work do have parallels in the Occitan text (see below).[109] As French linguist Claude Buridant has shown, the work of translation is multiple and includes commentary, explanation, and expansion of the source text.[110] Ilaria Zamuner has demonstrated, moreover, how the Occitan text's introductory lines are very different from this suggested Latin source.[111] The *Occitan Health Advice for the Layperson*'s discussions of bloodletting, moreover, are not found in johannes Hispaniensis's text.

The month-by-month presentation of advice (ll. 349–412) may well find its model in the short Latin prose works known as *Regimina XII mensium* (see above).[112] The following chart shows some of the correspondences between *Occitan Health Advice* and these regimens. As the parallels could be complicated by the variety of versions of the Latin texts,[113] I use only one calendar from one manuscript in what follows here.

Chart 3: Comparison of Occitan and Latin Monthly Calendars.

***Occitan Health Advice*'s Language**	***Regimina XII mensium* Advice**
Do not blood let in January;	do not do phlebotomy[114]
do not eat chard in February.	
In March you do not want to let blood, nor do cauterization, nor cupping. If then you have no great need for reason of illness [you will not want to consume] lentils or to eat sweet things.	Do not do phlebotomy[116] avoid sweet things[117]

109 See Suchier 1883: 531. Hunt 1999: 350, seems to imply that “translation” can be used for medieval vernacular works based on a combination of source materials.
110 Buridant 1983–1984, cited by Bérier 1988: 241.
111 Zamuner [Candiani] 2006: 173–176.
112 See Van Winter 2010.
113 See note 48. Note also Zamuner [Candiani] 2006: 172, and Pucci Donati 2007.
114 Fischer 2015: 116. Barbaud's example advises against bloodletting more generally, in the wintertime (Barbaud 1988: 336).
116 Barbaud 1988: 335; Fischer 2015: 115; Montpellier (Groenke Type I), Groenke 1986: 100, l. 13.
117 Fischer 2015: 116.

***Occitan Health Advice*'s Language**	***Regimina XII mensium* Advice**
Know that rue[115] would be good if you can drink it.	drink rue[118]
In April, one must avoid eating root [vegetables], but it is good medicine to blood let him who needs it.	Do not eat root [vegetables][119]
In May, one must not eat the head of anything, nor is it a good [time] for blood letting, but in this month, it makes good sense to consume fennel or absinthe.	Eat the head of no animal[120] consume fennel and absinthe[121]
In June one must often eat lettuce, and one must consume vinegar and all sour fruits at the table, when one is hungry.	Eat lettuce[122] drink vinegar[123]
In July it's not good to be bled nor to take medicine nor give it; as for ladies, it makes sense not to kiss them too often.	Do not do phlebotomy[124] abstain from coitus[125]
In August, he who is wise avoids eating, above all other things, things that the blood can mix and that do not create black bile, But, rather, light and comforting things: fresh meat, good breads and sauces of mint and of pennyroyal. These are of great benefit. In this month one must not buy beef or pork for eating.	 Avoid foods that generate black bile[126] drink pennyroyal[127]

115 One Montpellier version (Groenke Type I) also recommends rue for the month of July (Groenke 1986: 108, ll. 40–41).

118 Montpellier (Groenke Type I), Groenke 1986: 100, l. 15.

119 Montpellier (Groenke Type I), Groenke 1986: 102, l. 20.

120 Van Winter sees the prohibition of eating animal heads as stemming from Galen, Van Winter 2010: 107. One Montpellier version (Groenke Type I) adds that one should avoid the feet of animals as well; see Groenke 1986: 104, ll. 27–28.

121 Barbaud 1988: 335. Absinthe is often recommended for the month of May; see Barbaud 1986: 22. See Montpellier (Groenke Type I) Groenke 1986: 102, l. 22. Also Van Winter 2010: 106.

122 Fischer 2015: 116. See Montpellier (Groenke Type I) Groenke 1986: 106, l. 32.

123 Fischer 2015: 117, and Montpellier (Groenke Type I), Groenke 1986: 106, l. 32.

124 Barbaud 1988: 336; Fischer 2015: 119; Montpellier (Groenke Type I), Groenke 1986: 108, ll. 37–38.

125 Barbaud 1988: 336; Montpellier (Groenke Type I), Groenke 1986: 106, l. 36.

126 Montpellier (Groenke Type I), Groenke 1986: 108, ll. 43–44.

127 Montpellier (Groenke Type I), Groenke 1986: 108, l. 42. Pennyroyal is recommended in the month of January in other diet calendars, Barbaud 1986: 22.

***Occitan Health Advice*'s Language**	***Regimina XII mensium* Advice**
September is so lovely	
that it must make everyone joyful.	
For it does no man ill,	
if he has not been taken [ill] in August.	
Then one can eat without harm,	Eat what you want,[128]
all foods, better than in any other month of the year.	
and it is good to take medicine	
and to drink cow's milk.	drink milk[129]
In October one must consume	
sweet raisins, and one must press must.	Consume raisins and must[130]
Of all the [months of] the year, an empanada is not	
as healthy [as now], if it is well prepared,	
and one must use in sauces	
pepper, ginger and cloves.	consume pepper and cloves,[131]
In November do not bathe	Do not bathe[132]
if you want to protect yourself,	
but then are sufficiently beneficial	
bloodletting and cupping;	
and he is more skilled at composing,	
any man who wants to apply himself.	
In December a man must keep	
from eating cabbage morning and night,	
for these can then generate suffering	
which makes a man scream.	
But in this month he is very wealthy	
who uses lavender[133]	
often in his plate,	
and ginger, and very peppery [foods].	Use pepper[134]

In sum, *Occitan Health Advice for the Layperson* builds on the *Secretorum* tradition, incorporates a good deal of the *Regimen*'s thinking and formatting, and adds detailed medical and/or health information from other sources, making the Occitan work both a medical and a dietary text. The author's incorporation of a monthly calendar in his

128 Barbaud 1988: 337; Fischer 2015: 116; Montpellier (Groenke Type I) Groenke 1986: 110, l. 49. Montpellier (Groenke Type I), Groenke 1986: 110, l. 47–48.

129 Barbaud 1988: 335 and 337. Montpellier (Groenke Type I) specifically advises goat's or sheep's milk, Groenke 1986: 110, ll. 47–48.

130 Fischer 2015: 116; Montpellier (Groenke Type I) Groenke 1986: 112, l. 52.

131 Barbaud 1988: 335.

132 Montpellier (Groenke Type I) Groenke 1986: 112, l. 56.

133 The Latin text recommends parsley rather than lavender, Montpellier (Groenke Type I), Groenke 1986: 114, l. 62.

134 Montpellier (Groenke Type I) Groenke 1986: 114, l. 62.

vernacular work continues the informal Latin tradition of integrating calendars like these into other works, such that they are hidden and do not appear in book content lists (see above).[135] The number of close parallels between the Latin diet calendar and *Occitan Health Advice* – only the month of February shares no elements – leads me to think that the author may have had access to the medical volume housed in Avignon in the fourteenth century and not far from the proposed region of origin of our author.

Witnesses

We know of six Occitan witnesses to this *translatio*, four in verse and two in prose[136]:

B[137] Basel, Öffentliche Bibliothek der Universität Basel, D.II.11: a composite manuscript of mostly medical or related works, *Occitan Health Advice for the Layperson* is in a fourteenth–fifteenth century section, f. 163 verso (a fragment in prose tacked into a collection of pharmaceutical recipes),[138] Brunel manuscript number 355.[139] See Figure 2.

L1 London, British Library, Harley MS 7403: an Occitan miscellany in which this is the only medical text; end thirteenth century, ff. 49 recto–62 verso (complete in verse); Suchier's manuscript A, Brunel manuscript number 21.[140] See Figure 3.

L2 London, British Library, Additional 22636: a composite manuscript of medical works, thirteenth–fifteenth century, f. 46 verso (a three–line verse fragment appended to the *Thesaurus pauperum*); Suchier's manuscript C, Brunel manuscript number 15.[141]

S Sevilla, Biblioteca Capitular y Colombina, 5–5-20, f. 152 verso [f. 8 verso]: a composite manuscript of medical and other works; thirteenth–sixteenth century[142]

135 See the discussion of diet calendars, above, along with the related footnotes.

136 In agreement is Zamuner [Candiani] 2003: 746 and 2006: 170.

137 Though Foulet and Speer suggest assigning lower case sigla to fragments (Foulet and Speer 1979: 46), I prefer here to modify the sigla assigned by Zamuner 2003: 746; her sigla, B, L, L^1, V, V^1, and S are more informative than the A, B, C, and D assigned by Suchier in *Provenzalische Diätetik* (ed. Suchier 1894). I have adapted her sigla to make them even more explicit.

138 This fragment was first published in Wackernagel 1845. See the Jonas database online for a relatively complete description of the manuscript. For additional information, see Swisscollections, s. v. "Stephanus Aldebaldi; Rogerius Salernitanus; Benvenutus Grapheus Hierosolymitanus". See also Zamuner [Candiani] 2010: 216–217, who connects the Basel and Seville prose texts to the verse *Occitan Health Advice*.

139 Brunel 1935: 102.

140 Brunel 1935: 7.

141 Brunel 1935: 5.

142 See Zamuner [Candiani] 2010: 197–198, for a more detailed description of this manuscript.

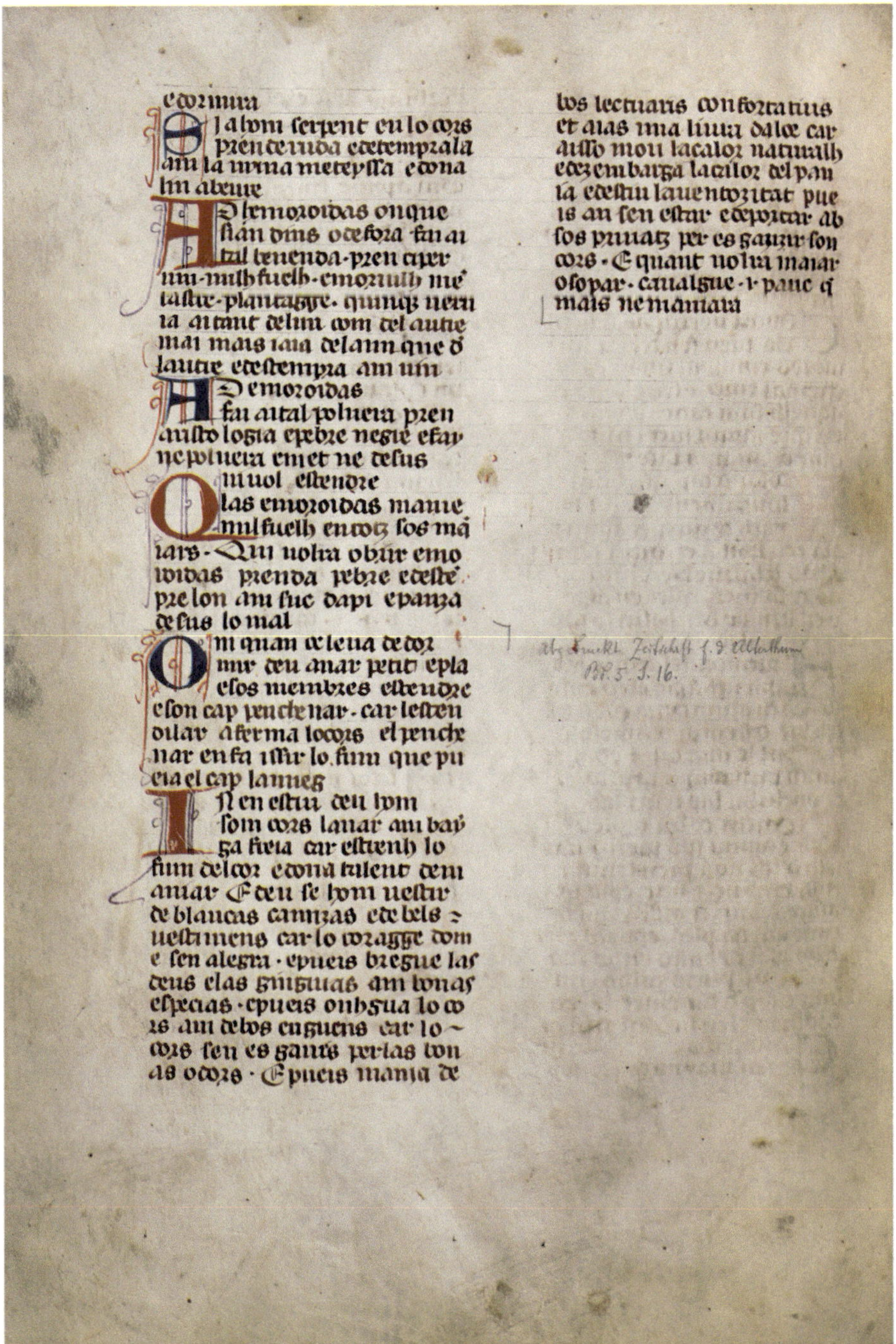

Fig. 2: Basel, Öffentliche Bibliothek der Universität Basel, D.II.11, f. 163 verso. With permission of the library.

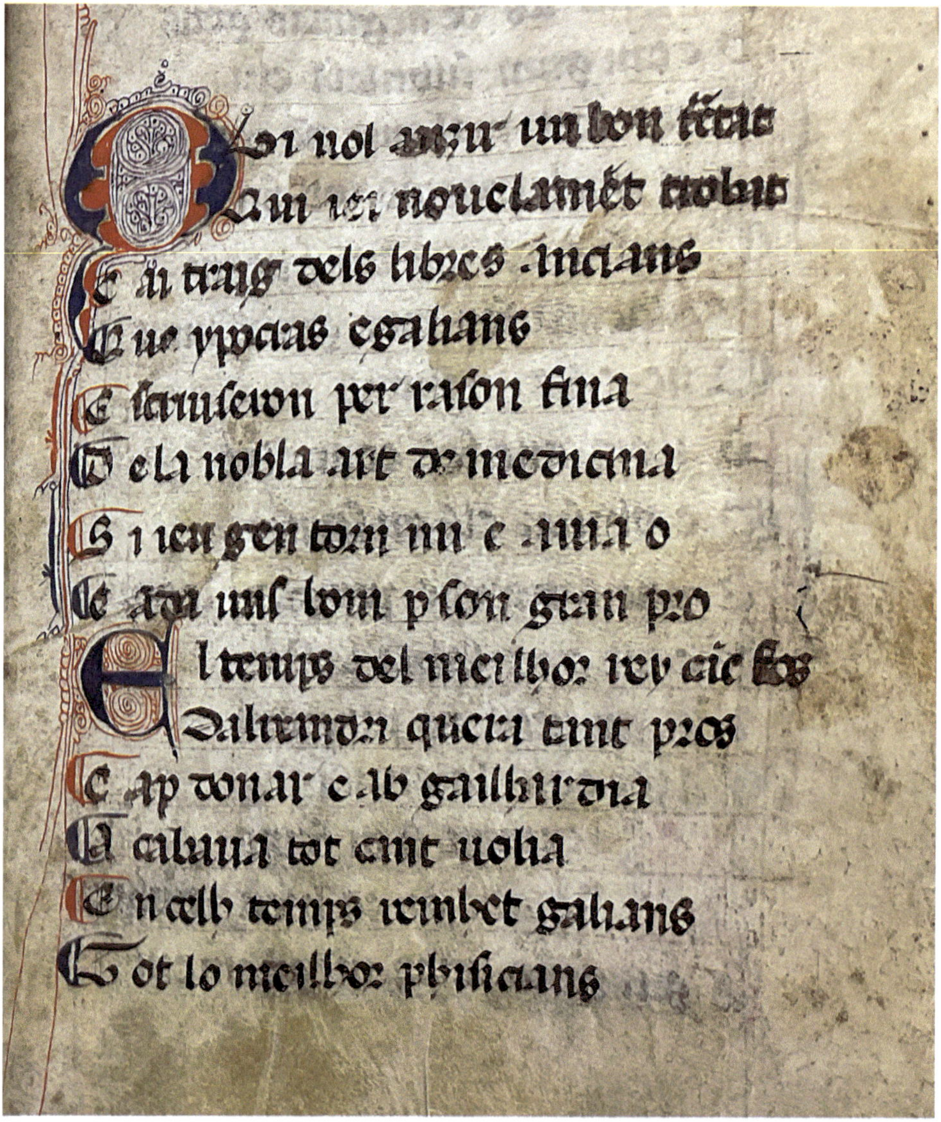

Fig. 3: Opening lines of *Occitan Health Advice*, British Library, Harley manuscript 7403, f. 49 recto. By permission of the library.

the prose fragment is in a section from the first half of the fourteenth century,[143] tacked into pharmaceutical recipes, as in B.

V1 Città del Vaticano, Biblioteca Apostolica Vaticana, *Barberiniani latini*, 311 (formerly Roma, Biblioteca Barberini X.129): a collection of health-related materials;[144] second quarter fifteenth century, ff. 19 recto–22 verso (complete verse text although shorter than the L1 witness; Suchier's manuscript D), Brunel manuscript number 327.[145]

V2 Città del Vaticano, Biblioteca Apostolica Vaticana, *Barberiniani latini*, 3574: a composite manuscript containing primarily seventeenth-century short works written mostly in French. The text is on a single fifteenth-century sheet, f. 84 recto–verso (fragment in verse; Suchier's manuscript B), bound with the volume, Brunel manuscript number 329.

Of these, the Harley manuscript, L1, contains the longest text, complete in itself. Vatican *Barberiniani latini*, 311, V1, is also complete albeit slightly shorter; it provides lines missing from the Harley witness. The second Barberini manuscript, V2, is a fragment with but 133 lines, though it provides a few lines not found in the other witnesses. There is no authorial attribution for the Occitan text in any of these witnesses; the oldest manuscript containing *Occitan Health Advice*, the Harley manuscript, is dated to the end of the thirteenth or beginning of the fourteenth century.

Suchier published an edition of the Occitan verse text in 1883 based largely on the Harley L1 manuscript; following Wilhelm Wackernagel's lead, Suchier assigned to the work the title by which the Occitan text is now known among food scholars.[146]

Italian philologist Yole Ruggieri (†1973), who published an edition of the *Barberinianus* 311 text, considered that it and the L1 manuscript should be considered as different redactions from a single Occitan exemplar.[147] She added that her Vatican witness is an independent redaction from the western area of Occitania close to Catalonia.[148]

Ilaria Zamuner argues that the two Vatican manuscripts and the Harley 7403 manuscript share a common source,[149] distinguishing these three verse witnesses from a version in prose which exists only in fragments incorporated into other works.

---

143 Zamuner [Candiani] 2006: 170.

144 See Zamuner [Candiani] 2004 for a detailed description of this manuscript, well-known for its illustrations.

145 Brunel 1935: 95.

146 The Jonas database online uses the Latin phrase *Epistola ad Alexandrum de dieta servanda* as the title for this work.

147 Ruggieri 1930: 203.

148 Ruggieri 1930: 203.

149 Zamuner [Candiani] 2005: 58.

That the prose fragments shift from second person to third in their instructions suggests to me a different pattern of transmission than that of the verse text.

Building on the observations of earlier scholars as well as my own, I propose this stemma for the longer verse witnesses:

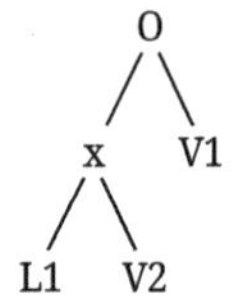

Fig. 4: Stemma of Manuscripts.

While the V2 witness is a fragment, containing roughly one-quarter of what I suggest was the original text, the V2 witness agrees with L1 against V1 in multiple instances. For example, l. 7, where L1 and V1 maintain the preposition "entorn", l. 10, where L1 and V2 both describe Alexander as "tant pros"; l. 13, where the two manuscripts evoke the period when Galen lived, against V1's description of "lo prous Galia"; l. 19, where V1's reading is short whereas L1 and V2 offer a *lectio difficilior*; l. 21, where L1 and V2 speak of the light of Galen's wisdom; ll. 25–26, missing in V1; l. 40, where L1 and V2 use "poizonava", against V1's "metzinava"; l. 47, where L1 and V2 mention a shirt, whereas V1 speaks of shoes; l. 55, where L1 and V2 speak of the bed and V1 speaks of the room; l. 65, where L1 and V2 use "confortar" against V1's "amundar"; l. 71, where only V1 evokes fennel; l. 72, where V1 speaks of pomegranate rather than bitter spices; l. 95, where L1 and V2 offer a *lectio difficilior*; l. 106, where V1 replaces "flowers" with "things"; l. 117, where V1 substitutes "cors" for "fortors"; l. 120, where L1 and V2 offer the more interesting verb "agusar"; l. 128, where V1 offers a quite different line; l. 131, where V1 offers a *lectio difficilior* against L1 and V2.

All three long witnesses have lines missing from each other, which suggests intermediary points of transmission from the source text. This edition attempts to recreate that source text by combining elements of the three versions.

Contents

The topics of *Occitan Health Advice for the Layperson* are as follows:

- Introduction and Presentation of Galen
- Reveille, with Subsections on Dress and Morning Medications
- Exercise and Conversation with Counsellors
- Meals and Table Manners
- Sleep Habits

- Entertainment[150]
- The Four Seasons, with subsections on What to Eat When, and When to Bloodlet
- The Months of the Year, with additional advice
- Conserving Body Heat
- Conclusion

The Prose Versions

The prose witnesses are so short that it is difficult to draw any conclusions regarding them. Ilaria Zamuner argues that two distinct translations of Johannes Hispaniensis developed in the Occitan language area, one in prose and another in verse.[151] What is certain is that a collection of pharmaceutical recipes originally from Montpellier (now Basel, Öffentliche Bibliothek der Universität, D.II.11)[152] includes at least two "recipes" in prose that repeat advice given in the *Occitan Health Advice* in verse;[153] these recipes are the prose tradition noted by Ilaria Zamuner.[154] The Seville fragment follows the same model, lines that repeat advice from the verse text inserted into a medical recipe.[155] French scholar Françoise Fery-Hue thinks that two distinct Occitan versions were translated during the fourteenth century, the shorter of which (I assume V2; she does not specify) was copied in or near Catalonia.[156]

---

150 The listing of entertainment activities, specifically games, may relate to what Jean-Michel Mehl calls "an invasion of play", "une invasion ludique", that begins in the thirteenth century (Mehl 2010: 42), and may help narrow the date of composition of *Occitan Health Advice.*

151 Zamuner [Candiani] 2003: 752.

152 "Recettes médicales": Basel, Universitätsbibliothek, D.II.11, ff. 153 verso–154 recto. See Zamuner [Candiani] forthcoming, for the most recent description of this manuscript. I thank her for sharing the article with me in advance of its publication.

153 Brunel 1957: 316–317, recipe numbers 120 and 121. Brunel 1957: 361n1, noted this coincidence in his edition of this recipe collection. I transcribe the Basel witness as follows: "Om quan se leva de dormir deu anar petit e pla e sos membres estendre e son cap penchenar, car l'estendilar aferma lo cors e.l penchenar en fa issir lo fum que pueia el cap la nueg. In en estiu deu hom som cors lavar amb ayga freia, car estrenh lo fum del cor e dona talent de manjar. E deu se hom vestir de blancas camizas e de bels vestimens, car lo coragge d'ome s'en alegra, e pueis bregue las dens e las gingivas am bonas especias e pueis onhgua lo cors am de bos enguens, car lo cors s'en esgauis per las bonas odors. E pueis manja de bos lectuaris confortatius et aias una liura d'aloe, car aisso mou la calor naturalh e dezembarga la calor del panja e destrui la ventozitat, pueis an s'en estar e deportar ab sos privatz per esgauzir son cors. E quant volra manjar o sopar, cavalgue .i. pauc, que mais ne manjara".

154 Galderisi 2011: 2.2.1254–1255, also considers the Basel prose manuscript as a fragment of the verse *Occitan Health Advice.*

155 Zamuner 2010: 217 transcribes this passage as follows: "Hon cant se leva de dormir deu anar petit e[m] pla, e sos membres estendre e lo cap penchenhar, [a]car l'estedilar aferma lo cors el pencenil ... lo fum que pueia el cap de nuet."

156 Fery-Hue 1992: 1369.

Dating the Text

Following Suchier, Zamuner notes that Matfre Ermengaud appears to cite the *Occitan Health Advice for the Layperson* in his *Breviari d'amor*, composed before 1323, which could set that year as the *terminus ante quem* for the composition of the medical text.[157] However, Peter T. Ricketts, recent editor of the *Breviari*, did not confirm this citation.[158] I do not see where, in the section suggested by Suchier, "De la natura dels quatre tems del an" (ll. 6415–6518), the *Breviari* presents any lines that repeat those of the *Occitan Health Advice*. What one might best think of as reminiscences are – at the most – common knowledge to a fourteenth-century learned author such as Matfre. Among other differences, the *Breviari* offers its readers much more temporal information (naming the astrological signs of the year, for example) than does *Occitan Health Advice*. I conclude that Matfre's direct source is not *Occitan Health Advice* and that we cannot use lines in the *Breviari* to date the anonymous text.

In order to date *Occitan Health Advice*, we must use the dating of its manuscripts, seen above, and linguistic information in the work. Note, though, that Zamuner dates *Occitan Health Advice* rather precisely to circa 1288.[159]

Linguistic Issues

In a text that is relatively short, we are not given a lot of data by which to consider the language or dialect of the translator/author separate from that of the scribe who copied the text.

Ruggieri's brief but detailed analysis of V1 included signaling orthographies expected in Catalan settings, such as *let* (her ll. 220, 331) for *laig* (ll. 277, 392); *fretura* (her l. 374) for *frachura* (l. 437); *profit* (her ll. 180, 321) for *profieg* (ll. 223, 380), along with the somewhat unusual subjunctive *vulla* (here l. 209); the passage of unaccented e to a, as in *malencolia* (here l. 296); the persistence of open o followed by /k/, as in *foch* (here l. 299); and the maintenance of l plus consonant, as in *malaltia* (her l. 370). She noted also in this witness a few French words (e.g. *par*, her l. 168, here l. 207). Despite these examples, the vast majority of linguistic details noted by Ruggieri keep this Barberini witness firmly in the Occitan linguistic area. The remarks below consider linguistic features of the text as presented in the Harley witness (L1), the text that provided the majority of the lines of this edition.

157 Suchier 1883: 529–530; Zamuner [Candiani] 2003: 757–758.

158 The suggested citation of *Occitan Health Advice* in Matfre Ermengaud's *Breviari 2* (ed. Ricketts 1989) occurs at lines 6489–6492. Ricketts identifies Galen for the reader, Matfre Ermengaud, *Breviari 2* (ed. Ricketts 1989: 265n5509), but makes no mention of a possible quotation from *Occitan Health Advice*. These lines were not discussed in Ricketts 1989.

159 See Zamuner [Candiani] 2006: 181, and 2003: 757–759.

The Language of the Author

The language of the author is most easily considered in the rhyme words, less likely to change when the text was copied.

- -n

We observe that the author probably did not pronounce final -*n* (examples: pargami < PERGAMENUS : enaissi < SĪC, 27–28; enaissi : mati < MĀTŪTĪNUS, 43–44; cisclatons : precios 81–82; molra : safra, 319–320. There is also that example of confortans : pans, 379–380, where we cannot know if the final letters were sounded or not. Suchier thinks this last rhyme points to the southeastern area of Occitan;[160] I think the “pans” orthography is not a specifically southeastern feature, as related spellings can be found in documents from many points west, including Dax, La Réole, and Bordeaux, as well as Narbonne and Montpellier.[161]

- *-tz*

The rhymes in the Harley manuscript suggest that the passage of final -tz to -s is in progress.[162] See eissamentz : dens, 61–62; auras : solatz, 139–140; palaitz : gais, 235–236; res : estaretz, 419–420; fenis : esperitiz, 425–426; semblantz : ans, 455–456. There are also twelve examples of the expected -tz : -tz rhyme.

- final -t not pronounced

The rhyme pair ferran : pezant, 191–192, suggests that for the author, the final -t was silent.

Individual Rhyme Words

In terms of individual rhyme words, a few merit additional comment.

- trazir, 19 < TRAĬCĔRE, “to swallow”, is a term associated with the region of Toulouse in the thirteenth century, the Limousin (traigir), Rouergue, and Aurillacois (tragir) in the modern period.[163]
- estendilatz, 46 < EXTENDERE. The expected spelling would be “estendilhatz” which would make a stronger rhyme with “reveilhatz” 45. Is this a matter of scribal error or representative of the author’s speech? I leave the question open.

---

160 Suchier 1883: 530, “den Südosten des Provenzalischen Gebietes”.
161 See Levy 1894–1924: 6.37b, s. v. pan, “bread”.
162 Ronjat 1930–1941: 2.283–284, §382.
163 Wartburg 1922–2002: s. v. TRAĬCĔRE, 13.2: 187; Mercadièr and Chapduelh 2023: s. v. avaler.

- oliver, 69, "olive tree", represents a Gascon spelling of the word,[164] seen as "olivier" in other dialects of Occitan (the "oliver" spelling is also found in Catalan).[165] Note that the author rhymed "oliver" with the name of another fruit tree, "presseguier", 70, which has the expected suffix for trees, -ier.[166]
- The rhyme pair of estieu : rieu, 97–98, carries a hint of western Occitan. The Gascon (or Vivaro-Alpine) spellings "estieu, estieus" are used through the text, as opposed to "estiu" found in other modern dialects.[167] The "rieu" spelling is more typical of Gascon dialects of Occitan.[168]
- The rhyme of freig : destreig, 327–328, seems limited to the regions of the Rouergue, Quercy and Toulousain.[169]
- The rhyme pair profeig : lieg, 221–222, is somewhat curious – *Occitan Health Advice for the Layperson* provides the only example of the "profeig" orthography in both sections of the *Concordance de l'occitan médiéval* 2005; this spelling does not occur in the *Trésor manuscrit de l'ancien occitan* 2017– (as of 5 September 2022). PROFECTUS should result in "profieg", the more usual spelling found at line 382. I call this orthography a curious scribal error but do not correct the text. I believe the author heard the /e/ clearly, as in "profechosa", l. 401, a word with a related etymon. We also have "naleig", 439 < NĔGLĒCTUS, not quite so rare; this spelling is used several times in the verse *Leys d'amors* (ll. 1944, 2185, 2191, 2201, 2821, and 6529).[170]
- pulment, 174, is not listed in the *Dictionnaire de l'occitan médiéval.* In that dictionary, the closest word in terms of spelling alone is "pumen, pimen" < PIGMENTUM, whose meaning of "spiced wine" makes no sense at this point in *Occitan Health Advice.* Polmen < PŬLMĔNTUM, "food, nourishment", is in the *Dictionnaire de l'occitan médiéval* and is the term intended here. The spelling with -u- instead of -o- seems typical of the Narbonne region.[171]
- janoier, 349. This orthography is found in the Montpellier *Thalamus* and in a cartulary from Albi.[172] The related spelling of "genoier" is found even more frequently in the *Thalamus*. One of the *Breviari* witnesses, Paris, Bibliothèque nationale de France, fonds français 858, uses "ginoier", see *Breviari 2,* line 3886 (ed.

164 Rei-Bèthvéder 2004: s. v. olivier.

165 Institut d'Estudis Catalans, *Diccionari*: s. v. oliver.

166 Adams 1913: 224–225.

167 Comité d'études gascon de la langue occitane 1998: s. v. été; Faure 2009: s. v. été.

168 Comité d'études gascon de la langue occitane 1998: s. v. rivière.

169 See Jud 1939: 207n1: "La zone de dretsa : metsa comprend grosso modo le Rouergue, le Quercy, le Toulousain."

170 Guilhem Molinier, *Las Flors del Gay Saber* (ed. Anglade 1926). Compare with the observation in Weisse 1883: §11.6.

171 See Levy 1894–1924: 6.433b, s. v. polmen.

172 *Trésor manuscrit de l'ancien occitan* 2017–, s. v. janoier and genoier; Levy 1894–1924: 4.104a, s. v. genier.

Ricketts 1989); this fourteenth-century manuscript has been localized to the Languedoc region.[173] The evidence suggests that this spelling can be localized to the area from Albi to Montpellier and west to Béziers.

The Language of the Scribe

The orthography of a vernacular text copied in the Middle Ages tends to represent the pronunciation and vocabulary of the scribe more than those of the author, as a scribe might take some liberties with his model but not change the rhymes. The features described below may be examples of Occitan developments from Latin generally; dialect-specific evolutions will be noted.

- A > e

The text demonstrates this development particularly in words based on ALĂCER +: alegrament 56, 475; alegramens, 79; alegrier 253, 459. The two examples of MANDŪCARE > menjar, 156, 172, are in lines taken from a Catalan-influenced witness. The Harley L1 scribe consistently spells this verb "manjar", 66, 158, 188, 190, 198, 204, 273, 294, 311, 329, etc.

- Open E > diphthong

The text offers several examples of diphthongization of etymological -E-: DELECTARE > delieg, 56; LĔCTUS > lieg, 55, 224, 231, 440; PŪLĒIUM > polieg 381; REX > reis 413, 414; rey, 28; SĔNIOR > seynher, 417.

- I > e

FRĪGIDUS > frejor 282, 306; freydor, 110;[174] *ĬNSĬGNARE > enseynhatz 146; MĔTĬPSE > meteis 400;[175] PĬSCIS > peis 288. There is also the example of MĬNUS > meins 291, a form Calvet describes as archaic.[176] In keeping with this movement are the examples of ĬPSEMENTE, ĬPSAMENTE > eissament and eissamens 61, 203, 246, 259, 345; however, consider ĬPSAMENTE > issament 457.[177] Van der Horst suggests that this orthography is first found in the Tarn-et-Garonne in the twelfth century and in the Hérault in the thirteenth.[178]

173 Brunel 1935: 43; it is Brunel manuscript 145.
174 Ronjat 1930–1941: 2.117, §280γ.
175 Pfister 1960: 82.
176 *Lo Rosari* (ed. Calvet 1997: xlviii).
177 Pfister 1960: 78–80.
178 Van der Horst 1981: 349; Pfister 1960: 78–80.

- Substitution of “i” for “e”

The only example concerns the proper name Alexander: Alixandri, 10, but compare Alexandri, 29, 413.

- I remains “i” and does not move to “u”

This feature marks the text as not from the Toulousain, where -I- moved to -u-. Examples include prima < PRĪMUS, 48, and variations on that root: primieiramens, 182; prim, 226; primiers < PRĪMARIUS, 239; primier, 460; and primaveira, 267.

- Orthography -ieu

François Zufferey observed that troubadour chansonnier R, from Toulouse, had as a dominant trait the -ieu orthography, representing the evolution of iu > ieu.[179] The forms in -ieu become dominant in the second half of the fourteenth century.[180] Examples in our text include: AESTĪVUS > estieu and estyeu, 63, 89, 97, 105, 265, 281, 312, 326; CAPTIVUS > caitieu 453; DEUS > Dieu or Dieus 417, 464; NĬX > nieus, 266; RĪVUS > rieu 98; VĔTŬLUS > vieu 422.

- Open O

Open O > o is a feature of western Occitan to Narbonne; in Catalan open O moves to “u”. Ronjat notes that -O- is maintained in western Languedocian and in Guyenne.[181] Our examples include FŎCUS > foch, 299 and the verb MŎVĒRE which presents the conjugated form moras, 217. The movement of open -O- to a -u- spelling can be seen in this text as well. This movement to -u- or to a diphthong is typical of Western Languedocian, from Narbonne to Catalonia.[182] Examples include: CŎQUUS > cuex, 176; LŎCUS > luecs, 480; NŎX or NŎCTEM > nuitz, 452; PŎSTEA > pueis or pueissas, 25, 137, 157, 197, 217, 228, 255, 423, and the third-person singular subjunctive form of the verb poder, puesca, 377.

- Closed/long O > u

I find one example: ŌSTIUM > vueis, 235.

- Orthography -iu

The Harley manuscript offers one example: CĪVĬTAS > ciutat, 262. The case of AESTĪVUS > estiu, 295, 297, comes from Barberini witness V1 with strong Catalan influence.

- U to “o”

The text offers at least two examples: BŬCCA > boca, 61, 68; FRUMENTUM > froment, 167.

---

179 Zufferey 1987: 108.
180 Compare *Lo Rosari* (ed. Calvet 1997: xlviii).
181 Ronjat 1930–1941: 1.119–121, §67.
182 See Zufferey 1987: 172, and Talfani 2022: 160–161.

- U > "u"

As Ronjat observed, the evolution of DUBITARE is somewhat exceptional; our text offers at least one sample of this treatment, duptava, 20.[183]

- Forms with -iei-

This spelling is localized to the region of Montpellier and Lodève in the modern period.[184] Examples include: primieiramens, 182; ribieiras, 479.

- Latin -ARIUS, -ARIA > -ier, -eira

The -eira development is localized to a swath that runs from the areas near Nîmes and the Cévennes, west towards Montpellier, Foix, Toulouse, and Narbonne. The text offers these examples: carreiras < *CARRARIA, 127; ribieiras < *RĪPARIA, 479; primieiramens < PRĪMARIUS, 182. Compare with escudiers, 161 < SCŪTĀRIUS; cavalier, 133, 241 < CABALLARIUS; messatgiers, 134; presseguier, 70 < PĔRSĬCUM + -ARIUM. This text also offers the examples of culheirat, 86 < COCHLEAR + -at and primaveira < VER, 267.

- Latin -BS-

The text offers aussens, 366 < ABSINTHIUM. Pfister observes that forms with au- are found only west of the Rhône.[185] Ronjat places "aussens" in dialects of Rouergue, Nîmes, the Cévennes, Quercy, and the Landes – generally the western dialects of Occitan.[186]

- Latin -BT-

The text offers one possible example: MALE HABITUS and related words > malaute, 18; malautes, 440; malautia, 35; malautia, 433. Ronjat reports that "malau(t)" is found east of the Rhône and also in the Velay, in Narbonne, Carcassonne, the Lauragais, Toulouse, Agen, and Aquitaine.[187]

- C + A stays /k/

The examples of C + A remaining as /k/, initially and internally, are multiple: blanca, 48; boca, 61, 68; cabra, 277; cabridet, 284; calor and calors, 65, 206, 281, 326, 418; cambra, 57; camis, 127; camisa, 47; cansoneta, 247; cant, 129; canut, 191; cap, 49, 363; capons, 334; carbons, 333; carbonadas, 336; carn and carns, 283, 288, 380, 383; carnos, 78; carreiras, 127; cassada, 258; castel, 262; castiar, 195; causa, 113; cavaliers, 133; cavalcaras, 126; escarpir, 50.

---

183 Ronjat 1930–1941: 1.267–268, §148.

184 *Lo Rosari* (ed. Calvet 1997: xlix), quoting Ronjat 1930–1941: 1.197–198, §113.

185 Pfister 1960: 37.

186 Ronjat 1930–1941: 2.169–171, §310.

187 Ronjat 1930–1941: 2.167–168, §309δ.

- -CT- > /tʃ/, represented by -ch-

The text offers several examples, including FRACTURA > frachura, 437; LACTŪCA[188] > laychuguetas, 278 and lachugas, 368; ŎCTŌBER > oichuvre, 393 (this spelling is not included in the *Dictionnaire de l'occitan médiéval*, s. v. octobre);[189] PĔCTĬNARE > penchenhar, 49. Other examples include PROFECTUS > profechosa, 401; DICTUM > dich, 310; along with adonchas, 289, 311, 382, 401; foch, 299; richs, 82. Ronjat suggests this orthography and pronunciation extend from eastern Occitania to Narbonne, Lodève, Guyenne, Auvergne and the southern Limousin.[190]

- Palatalization of -LL- and -L-

Zufferey called palatalization of -LL- very typical of chansonnier C, from Narbonne, and characteristic of Languedocian near Catalonia, a trait that "probably did not go beyond the regions of Carcassonne and Narbonne".[191] Examples in this text include: arbilha, 67 < ARVĪLLA; belhas, 272 < BELLUS; celh, 13 < ĬLLE; novelhas, 135 < NŎVĔLLUS; nulha, 147; nuilh 387 < NŪLLUS. We also have examples of palatalization of a single -L-. The example of abrilh, 359 < *APRĪLIU has received special consideration: the *Dictionnaire de l'occitan médiéval*, s. v. abrilh, says "this form in -[ʎ] belongs mostly to the northern area of gallo-romance".[192] Ronjat distinguishes between dialects that vocalize the "l" and those in the southwest that evolve towards palatalization and "-lh",[193] the case here, as with FENŬCŬLUM > fenolh, 366; JŪLIUS > julh, 371; and ŎCŬLUS > vuelhs, 232. The example of MELIUS > miels, 74, 251, 252, 389, is found by Ronjat in Nîmes, the Cévennes, Gévaudan, and the Aurillac region.[194]

- Nasal + "p" or "b"

Several developments are possible. Sometimes, we find a -p- or -b- inserted to assist pronunciation: MĔMŎRE > menbre, 99, 415. The case of AUTUMNUS > automp, 266, is different; the addition of "-p" may be an effort at a learnèd spelling or for reinforcement (the following word starts with a vowel).[195] There is also the example of DAMNARE and its derivative dampna, 433, where the internal -p- is not etymological and

---

188 See Ronjat 1930–1941: 2. 171–175, §311. Ronjat points specifically to Provençal "lachugo", Béarnais "leituge", 2:11.

189 See also Pfister 1960: 113. We find "ochovres" in *Flamenca* (ed. Gschwind 1976: l. 6658), "ochoire" in Agen in 1242, "oytoyre" in Olliergues (Puy-de-Dôme) in 1252, "octoire" in Millau in 1364, "otoyre" in Narbonne in 1380, suggesting that the spelling in our text is more to the west than to the east. See Wartburg 1922–2002: s. v. OCTOBER, 7: 308b.

190 Ronjat 1930–1941: 2. 171–175, §311.

191 Zufferey 1987: 148, "ne devait guère déborder le Carcassès et le Narbonnais", citing Ronjat 1930–1941: 2.148–150, §303.

192 *Dictionnaire de l'occitan médiéval*, s. v. abrilh, says "ce type en -[ʎ] appartient surtout au nord du gallo-rom.", pointing to Wartburg 1922–2002: s. v. APRĪLIUS, 25: 59a–60a),.

193 Ronjat 1930–1941: 2.305–311, §394.

194 Ronjat 1930–1941: 2.322–323, §401.

195 Zufferey 1987: 146.

may again be added either for reinforcement or to appear learnèd. Latin TĔMPUS always maintains the "-p" in this text: temps, 9, 13, 33, 88, 90, 104, 106, 192, 222, 294, 300. This may be a learnèd silent insertion or used to distinguish the noun from a potential homonym. The internal -p- in destemprament, 421; temporas, 263; temprada, 304; tempratz 267, 273, 480, does not surprise.

- Latin -MBR- maintained

As proper nouns, setembre, 302, setempbres, 385; novembre, 399; decembres, 405, are somewhat exceptional cases. Ronjat suggests the -mbr- spelling shows influence of French;[196] a learnèd spelling cannot be discounted either. Zufferey suggests the -p- in setempbres, 385, was added for reinforcement.[197] Compare menbre, 99, 415; membre, 310, comes from Catalan-influenced witness, V1.

- PR > br

We have the clear example of abrilh, 359 < APRILE.[198]

- Examples of -x- orthography

The maintenance of the Latinate spelling in dextre, 227 < DEXTER, may be an attempt at a learnèd spelling.[199] The orthography of cuex, 176 < CŎQUUS, suggests geographic proximity to Catalan. Other examples of -x- in this edition come from the Catalan-influenced witness V1: axi, 300; exida, 309; exiras, 57; exits, 308; mateix, 209.

- Rhotacism

I find only one example: FRAGRARE > flairar, 102, 435 and flayran, 184.

- Metathesis of R

Again, I find only one example: PĔRSĬCUM > presseguier, 70.

- Treatment of Latin -PT-

We have the example of caitieu, 453 < CAPTIVUS. Ronjat suggests this orthography is Languedocian and Aquitainian.[200] Other examples show a different approach to late Latin -PT-: Ronjat pointed to the evolution of DUBITARE as exceptional for several reasons;[201] duptava, 20, follows his examples. Comptats, 269, is from Catalan-influenced witness V1.

---

196 Ronjat 1930–1941: 2.234–236, §350.
197 Zufferey 1987: 146.
198 Ronjat 1930–1941: 2.232–234, §349.
199 Consider Ronjat 1930–1941: 2.190–192, §321.
200 Ronjat 1930–1941: 2.165–169, §309.
201 Ronjat 1930–1941: 1.267–268, §148.

- Final "-n"

Final "-n" is generally maintained, seen in the many examples of "non" (23 times) as opposed to the fewer examples of "no" (a total of 4 examples,[202] two of which come from Catalan-influenced witness V1). We also have rason, 5, 189 and rescon, 341, where "rescos" is generally the more frequent orthography. There are also examples of mobile "-n", seen clearly in some rhyme pairs : mati : enaissi, 43–44; camis : jardis, 127–128; afans : grans, 151–152; sa : t'aura, 169–170; atressi : mati, 193–194; coichos : digestios, 197–198. Note the orthography of "matin" at lines 406 and 482.

- Orthography of /ɲ/

The text offers several spellings to represent the sound of /ɲ/, the expected Occitan "-nh-" and also "-gn-". This latter is seen in regne, 144, and the verb forms relating to bloodletting: sagnar, 351; sangnia, 361, sangnar, 371, sangnia, 402. Zufferey thinks the "-gn-" spelling is Latinate.[203] We also have the expected "-nh-" as in banhar, 399; enseynhatz, 146, and ensenhatz, 413; fenholh, 366; reinhet, 13; penchenhar, 49; poinhes, 343; seynher, 417. Lastly, there is the case of jung, 367, and sueng, 458, with "-ng" in place of the expected "-nh".

- Orthography of /j/

The expected orthography of "-lh-" is frequent: arbilha, 67; culheirat, 86; failha, 37; gailhardia, 11; meilhor(s) and melhor 9, 14, 170, 238, 254; reveilhatz, 45; vuelhs 232; vuelhas, 203, 239, 351. However, these letters are also found in places where the pronunciation would be /l/: belhas, 272; novelhas 135, 380 against novelas, 248; novelament, 2, novel, 284; nulha, 147.

- -sc- > ss

The examples do not surprise: CRĒSCĔRE > creisser, 122; EXPERGISCERE > espreisser, 234; JŪSCĔLLUM > jussel(s) 319, 397.

- -tz = /s/

We have the expected reduction, seen in rhymes: eissamentz : dens, 61–62, auras : solatz, 139–140, palaitz : gais, 235–236, res : estaretz, 419–420, fenis : esperitz, 425–426, semblantz : ans, 455–456.

- Final "-t" retained

Retention of "-t" is a feature of Languedocian, perhaps particularly of the area near Montpellier.[204] Examples include: aitant, 420, 422; cant, 12, 26, 88, 106, 129, 160, 188, 202, 233, 326, 370, 429; enant, 192; quant, 45; tant, 10, 18, 21, 176, 194, 385. Compare tan, 17, 344.

---

202 The relevant lines in the Harley witness are 338 and 398.

203 Zufferey 1987: 147.

204 See Zufferey 1987: 181.

Morphology

The *Occitan Health Advice for the Layperson* represents Occitan case usage of its period; the two-case system of the twelfth century has fallen out of use, though a few traces of it can be found, for example, in line 14, with its mix of case endings: "tot lo meilhor phisicians". Other examples include "bona odors" (118, 119), "ab vin que sia clars" 183. The "-s" added to "estieus" at line 265 may have been to create an eye rhyme. As a general rule, though, "-s" is used to mark plurals, rather than case.

Verbs

The vast majority of verbs used in the *Occitan Health Advice* follow expected rules of conjugation. However, some cases warrant notice. The verb "voler" presents with examples of subjunctive forms that suggest Gascon influence: specifically volh' 404 and vulla, 209, though this latter is from Catalan-influenced witness V1. The verb "servir" seems to be conjugated on the simpler Gascon, rather than Languedocian model: see serva, 76 and serves, 177. Most distinctly Gascon is the text's use of the synonym for "annar", "ir", which exists only in the Gascon dialect, iras, 159. Our text offers examples of the third-person singular preterite verb ending "-et", as opposed to the "-ec" form limited to the regions of Toulouse and the Rouergue: atrobet, 23; enseynet, 24; reinhet, 13.[205] Our text also presents several examples of the third-person plural verb ending "-on": annon, 264; escriuseron, 5; podon, 407. This ending is characteristic of a large region that extends from the Auvergne and Quercy to the Gévaudan and the area near Nîmes.[206]

Regionally Specific Vocabulary

- Calendor, 109, is a somewhat rare word, according to Jud, found only to the east of a line linking Aurillac and Narbonne.[207] The "kalendas" spelling with initial "k-", 279, 280, 301, suggests consciousness by the V1 Barberini scribe of a semi-learnèd source.[208] "Calendor", with an initial "c-", suggests a word that has fully entered the language.
- Jaser, 449, "bed", is a another somewhat rare word in terms of this meaning. "Jaser" is more often seen as an infinitive, "to lie down, to go to bed". Wartburg 1922–2002: s.v. JĂCĒRE, 5: 1, points to the Old Occitan noun "jazer", "housing",

205 See Talfani 2022: 159.
206 Jud 1939: 209.
207 Jud 1939: 204.
208 See Jud 1934: 16n1.

"hébergement". The example in Raynouard's *Lexique*, s. v. jasier,[209] is taken from Paris manuscript Bibliothèque nationale de France, fonds français 2180, manuscript P of *Girart de Roussillon*, whose language appears to be Occitan "of the northwest, Perigord or Quercy";[210] the use of "jaser" as a noun may be a western Occitan regionalism, familiar beyond the Perigord and Quercy. My efforts to find the term in modern western Occitan dialect dictionaries have been unsuccessful: Jean Daniel offers "juc", "bed";[211] Aimé Vayssier comes closer to the usage in *Occitan Health Advice*, s. v. jas, "gîte, couche d'un animal" or the unexposed side of a melon.[212]

Other Vocabulary Considerations

- Verderos, 446, has been called a thirteenth-century hapax legomenon by Wartburg 1922–2002, a statement seconded by the *Dictionnaire de l'occitan médiéval*;[213] the only example of this orthography, i.e. of this word, is taken from this text. Note that V1 reads "verdegos". Levy cites just these two examples.[214] Both forms connect to the color adjective "green". Adams calls the "-os" suffix "exceedingly common",[215] adding that words with "-os" formed on adjectives have "hardly any change in meaning from that of the single word";[216] he does not include "verderos" in his list of words with the "-os" suffix. Levy connects the word to Catalan "vèrdic", "grünsauer" ("acid green", adjective), "vérdich, -ca", and to the Spanish "vérdica", "Grünsäure" ("acid green", noun). A better comparison is to the Catalan "verdós, verdosa", "green, unripe".[217]

Linguistic Conclusions

From the data analyzed here, it appears evident that both the author of *Occitan Health Advice for the Layperson* and the scribe who copied the text into the Harley L1 manuscript were from the western dialect area of Occitan. We can eliminate Toulouse as a point of origin for both of them. The areas near Nîmes, Montpellier or

209 Raynouard 1836–1845: 3.583
210 *Girart de Roussillon* (ed. Hackett 1953–1955: 3.466).
211 Daniel 1914: s. v. lit.
212 Vayssier 1878: s. v. jas.
213 Wartburg 1922–2002: s. v. vĭrĭdis, 14: 509b; *Dictionnaire de l'occitan médiéval*, s. v. verduros.
214 Levy 1894–1924: 8.663a, s. v. verduros.
215 Adams 1913: 324.
216 Adams 1913: 328.
217 *Gran Diccionari de la llengua catalana*, s. v. verdós, verdosa.

Narbonne are plausible loci for both the author and the scribe.[218] Furthermore, the use of Latinate spellings suggests that the author and/or the scribe sought to impress the reader by using these forms,[219] adding to the *auctoritas* and medical trustworthiness of the work.

Versification

Let us remember that the Occitan author of this *Health Advice for the Layperson* used as the framework of his text a work in Latin prose, which he translated into Occitan verse. The Harley manuscript presents a text in octosyllabic rhyming couplets, with relatively few imperfect rhymes. We find assonance at confortar : culheirat, ll. 85–86; fa : a, ll. 387–388; penre : beure, ll. 391–392. Only one rhyme pair exceeds the syllabic minimum, the potentially rich rhyme of del lieg : delieg, ll. 55–56.

The author has a certain poetic skill, seen also in his use of enjambment, lines that are not end stopped (see ll. 67–68, 228–229, 363–364, 389–390, 393–394, and 405–406). A less talented poet would compose eight-syllable lines in which the complete thought stayed in the single line.

One figure of rhetoric used well by the author is repetition or anaphora, most visibly toward the end of his work. As Nathaniel B. Smith observed, "any expressive repetition of words or word groups ... produces a reinforced effect greater than the sum of its parts".[220] In this vein, repeating an initial word, even if it itself has minimal meaning, is significant. Consider the series of lines that begin with the word "e", "and", ll. 121–124, 139–145, and 316–318 or the word "ab", "with", l. 442–454. The author could have varied his vocabulary but chose not to, emphasizing things that work together. The use of anaphora is particularly strong at the end of the Occitan text: lest the reader not recognize that "gaug", "joy", is the key to health, the author repeats the word no fewer than thirteen times in the last lines of his text. Calling attention to this emotion, the author alerts the reader first at line 459 with "ab gran gaug", beginning some ten consecutive lines with the phrase "ab gaug" (ll. 459–474). "Gaug" continues at lines 475 and 482. The message could not be more clear.

Thus, while not a great poet, the anonymous author did have some literary talent with which to present his medical and culinary advice.

---

218 Brunel thought the manuscript's origin was in Narbonne, Brunel 1935: 7.
219 See Lafont 1983: 35, for a brief discussion of the use of Latinate vocabulary by the translator of the *Albucasis*. Lafont suggests that the Occitan *Albucasis* or *La Chirurgie d'Albucasis* uses Latinisms for reasons different that what I propose here.
220 Smith 1976: 71.

Vocabulary of Specific Interest

(Words marked with an asterisk in the translation refer to this listing).

Afan (l. 407). Bondurand translates afan as "apostèmes", "tumors";[221] no such translation is offered by the *Dictionnaire de l'occitan médiéval*, s. v. afan; Pansier offers the translation "travail, gain" ("work, labor, profit")[222] which suggests a meaning of "suffering" in the context of *Occitan Health Advice for the Layperson.*

Agriam (l. 369). The precise meaning of this plural noun has been the topic of some discussion.[223] Suchier and Levy translate "agriam" as sour fruit, "saure Frucht".[224] Adams cites "agriam" as "new word" in Occitan with collective force.[225] Citrus fruits were certainly consumed in medieval Occitania, lemons serving as one marker of southern French cuisine.[226] The Occitan-area recipe collection, *De modus viaticorum preparandorum et salsarum* (a title often shortened to the *Modus*; see Culinary Context, below) includes a recipe for "limonieyra blanca" which clearly calls for lemon juice (its recipe number 41).[227] Citrus fruits, specifically oranges, lemons, and grapefruit, were among those recommended by Arnaldus de Villanova as mildly laxative.[228]

Assa (a variant reading at l. 402). Wartburg 1922–2002: s. v. ASA, points the reader to asafetida, the gum resin of the root of a plant used in medicine. Wartburg 1922–2002 suggests *Occitan Health Advice*'s use, if indeed, is a fourteenth-century hapax legomenon,[229] though the Occitan is used by Abraham Avigdor (1351–1402) in his translation into Hebrew of Gerardus de Solo's *Practica super nono Almansoris*.[230] The Florence herbal describes asafetida in these terms:

---

221 Bondurand 1895: 204.

222 Pansier 1924–1932: 5.159, s. v. afan. We must note that "travail" in French, "labor" in English, are the terms used to describe the efforts of a woman during childbirth.

223 The *Dictionnaire de l'occitan médiéval*, s. v. agrum, says the noun is derived from "agre" with the collective suffix "-um"; Wartburg 1922–2002: s. v. ACER, 24: 96b.

224 Levy cites Suchier who connected "agrium" to "agrum, eigrun", "sour fruit", Suchier 1883: 639, Glossary s. v. agriam.

225 Adams 1913: 82.

226 Lambert 2002: 70; Hyman and Hyman 1983: 70. See also Adamson 2004: 23.

227 This text is discussed in further detail below.

228 Carré 2019: 179; Arnaldus de Villanova, *Regimen sanitates ad regem Aragonum*, chapter 10: "De usu fructuum" (ed. García-Ballester and McVaugh 1996: 442–448).

229 Wartburg 1922–2002: s. v. ASA, 25: 402.

230 Bos, Hajek and Mensching recognize that as(s)afetida could be Latin or Occitan; they lean towards the vernacular, transcribing Abraham's word as "'Š' PYṬYD'", see Bos et al. 2021: 352.

Note: Unfortunately, the *Dictionary of Old Occitan Medico-Botanical Terminology* online (*DiTMAO*) was not available for consultation when this book was completed.

> Asa fetida es cauda e seca al .iiii. gra. E es goma d'albre que nais otra mar. E val contra la suffocatio matricis e contra dolor de dens e als asmatix.[231]
>
> (Asafetida is hot and dry to the fourth degree. And it is a tree gum that comes from overseas. It is useful against suffocation of the matrix and against toothache and for asthmatics.)

The *Thesaur de pauvres*, an Occitan text now in Chantilly (Musée Condé 330), offers a synonym for the resin, "apesamen".[232] One Auch pharmaceutical collection from circa 1441 uses it in a concoction for a pain in the side.[233]

Aste de porc (l. 336). In the kitchen, an "ast" is a spit or skewer;[234] by extension, an "aste“ is a piece of meat cooked on a spit or served on a skewer, in this case a piece of pork (see "raust and raustir", below). This mode of cooking and presentation is suggested in other Occitan documents, including the *Roman d'Esther*,[235] and illustrated twice in one manuscript of the *Jaufre* romance (Bibliothèque nationale de France, fonds français 2164, ff. 77 recto and 110 recto).

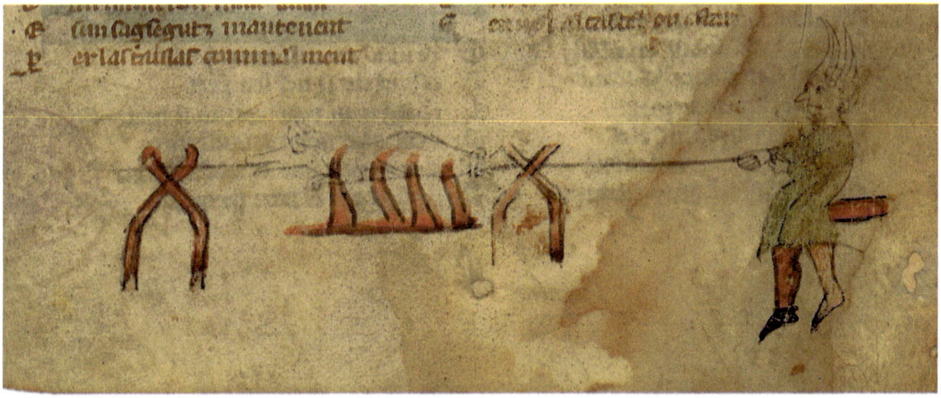

Fig. 5: Meat roasting on a spit, *Roman de Jaufre*, Bibliothèque nationale de France, fonds français, 2164, f. 110 recto. Source: gallica.bnf.fr.

231 "Aloes es caut e sec" (ed. Milani: 377).

232 Corradini Bozzi 1997: 276.

233 Recipe 95, "Costat. A dolor de costat, pren grana de ly e de ansens e de juelh, de cascuna p*er* engals partz, e met assa en *un* saquet fagz del gran la hon es la dolor, e apres sia mes bolir lo sac am de vy pur blanc, e quant sera bolhit, sia premit e mes tot caut sobre lo mal, e hostara la dolor" ("The side. For pain in the side [or the ribs], take a bit [seeds] of dregs of wine and of absinthe and of tares, equal parts of each, and put asafetida in a pouch made of grain there where the pain is, and then, boil the pouch with pure white wine, and when it has boiled, it will be ready and place it hot on the sore [spot] and it will remove the pain"), Corradini Bozzi 1997: 249. I observe that "assa" in this passage could mean something very different: "these" or "enough"; the word is not included in the glossary, see Corradini Bozzi 1997: 422.

234 Levy 1894–1924: 1.92a, s. v. ast.

235 *Roman d'Esther* (ed. and transl. Thiolier and Notz-Grob 1997: l. 143).

Aucels volans (l. 318). "Flying birds" is a large, general term; I understand the author to mean wild as opposed to domesticated fowl. A medieval noble table could see any number of birds as part of the menu, from peacock to partridge to heron;[236] a peasant could also eat wild fowl, such as duck, partridge or quail (see l. 275). If we assume wild fowl are captured during a migratory season, then autumn would be the recommended period, when the birds are at their fattest.

Aussens (l. 366). Wormwood (*Artemisia absinthium*), one of several plants used in the modern drink absinthe, is a bitter aromatic incorporated into any number of medieval pharmaceutical recipes.[237] The *Elucidari* devotes a paragraph to the plant, describing it as warm and dry and very bitter, with multiple medicinal properties. Because of its bitter taste, the *Elucidari* does not recommend consuming it, "sia desplazent a la boca per sa agudeza et amaror" ("it is displeasing to the mouth because of its sharpness and bitterness"),[238] and does not associate wormwood with any specific season. Arnaldus de Villanova describes the many beneficial properties of "absinthe wine", wine poured over wormwood leaves, so that the flavor and properties of the plant have passed to the wine.[239] Patrick Gifreu, translator of Arnaldus's *Livre des vins*, adds that one should harvest wormwood in June or July, before the flowers have blossomed, timing that aligns with the recommendation of our text.[240] Absinthe is called for in three different recipes in the Avignon pharmaceutical recipe book[241] and in two recipes in the Montpellier collection.[242]

Autupne (l. 303). Ibn Butlan describes autumn as a "balanced" season, "Sa nature est modérément froide et sèche; la sécheresse est l'effet de la chaleur de l'été tandis que le froid vient de la proximité de l'hiver" ("Its nature is moderately cold and dry; the dryness is the effect of summer's heat while the cold comes from the proximity of winter"),[243] concepts picked up by *Occitan Health Advice for the Layperson*. Given the season's cool and dry qualities, humoral theory would advise, as does this text, consumption of foods that are moist and warm. The *Elucidari* takes a somewhat different approach to the season, agreeing that it is cold and dry, but saying that it is unstable and harmful, "es may pestilent qu'els autres temps per razo de sa granda mutabilitat" (It is more pestilential than other seasons on account of its great changeability).[244]

---

236 Adamson 2004: 37–39.

237 Brunel 1957: 302–303 and 306, recipe numbers 12 and 30; 1966: 512–513 and 514, recipe numbers 12 and 18, among others; "A Fourteenth-century Provençal Receptary": recipe number xxxiiij 2 and page 216. Somewhat curiously, the plant was not part of the Bonis brothers pharmacopoeia.

238 *Elucidari* (ed. Hershon, Ricketts et al. 2018: 425).

239 Arnaut de Villeneuve, *Le Livre de vins*, p. 155.

240 Patrick Gifreu in Arnaldus de Villanova, *Le Livre des vins* (transl. Gifreu 2011: 154n134).

241 Brunel 1966: 512–513, 514, recipe numbers 9, 12, and 18.

242 Brunel 1957: 302–303 and 306, recipe numbers 12 and 30.

243 Ibn Butlan, *Taqwim al-Sihha* (ed. Elkhadem 1990: 227).

244 *Elucidari* (ed. Hershon, Ricketts et al. 2018: 262).

Fig. 6: Absinthe, *Tacuinum sanitatis*, Bibliothèque nationale de France, nouvelle acquisition latine 1673, f. 33 verso. Source: gallica.bnf.fr.

Baladas (l. 248). Medieval Occitan authors created and exploited a large number of different poetic genres, among them the "balada", whose name, derived from the verb "to dance",[245] implies a dance form and/or a song to accompany dancers. The *Canso de la crosada* includes baladas as one of the entertainments mentioned, making clear that baladas were sung.[246] Often translated as "ballad,"[247] the term is not included in the *Leys d'amors* nor other medieval guides to poetic composition, which is somewhat surprising. Pierre Bec would consider the balada as a popular form, which would explain its absence from the contemporary theoretical literature, though he has difficulty explaining why troubadour Pons de Capduoill composed in the genre.[248]

Bledas (l. 350). Although chard is recommended for consumption in winter by Ibn Butlan,[249] *Occitan Health Advice* suggests the contrary. The *Elucidari* recognizes chard as a common vegetable with several medicinal uses, within limits, "trop uzada, malas humors engendra"[250] ("overused, it engenders bad humors"). The Montpellier medical recipe collection includes chard as a remedy for tenesmus;[251] the fourteenth-century Princeton recipe collection (Princeton University Library, Garrett 80) incorporates chard in a prescription for a skin condition.[252]

Cabridet (l. 284). The reference to a "good, young goat" is intiguing.[253] The author insists on the small size of the animal, for "cabridet" is a doubled diminutive: cabra > cabret/cabrit > cabridet.[254] While goat meat is no longer a major source of protein in the United States, we know that goat meat was sold by butchers in medieval Montpellier,[255] Albi,[256] and other Occitan cities as well. The *Elucidari* sings the praises of goats:

245 Wartburg 1922–2002: s. v. BALLARE, 1: 218a.

246 The term "balada" occurs three times in the *Canso de la crosada*, at laisse 158, line 31: "que cascus ditz balada e verset o canso", at laisse 203, line 107: "que cantan las baladas e cansos e vercetz", and at laisse 213, line 110: "que cantan las baladas e los versetz leugers" (ed. and transl. Martin-Chabot 1972: 2.118, 3.182, 3.300). These last two citations clearly connect the verb "cantar" "to sing" with "baladas".

247 Raynouard 1836–1845: 2.174a, s. v. balada.

248 Bec 1977–1978: 1.229–230.

249 Ibn Butlan, *Taqwim al-Sihha* (ed. Elkhadem 1990: 17).

250 *Elucidari* (ed. Hershon, Ricketts et al. 2018: 427).

251 Brunel 1957: 299, recipe number 1. Tenesmus, "tenasmon" in this text, is a clinical symptom, feeling the need to pass stools, despite an empty colon.

252 "Per berbeze", Corradini Bozzi 1997: 135. "Berbeza, berbeze" is an otherwise unknown word; I trust Corradini Bozzi's translation, "malattia cutanea", see Corradini Bozzi 1997: 424 Glossario s. v. berbeza.

253 See Wartburg 1922–2002: s. v. CAPRA, 2: 295b.

254 Adams uses "cabridet" as his example, Adams 1913: 239.

255 City of Montpellier legislation spoke to sales of meat that came from healthy animals: "En ulh [sic] luoc del mazel non sia venduda carn de boc o de cabra, ni carn de moria o enferma o lebroza, o de bestia que vivens non vuelha manjar; ni negun non venda carn de feda o daret per moton crestat, ni carn de trueia per carn de porc" ("In no place of the meat market shall be sold meat of cattle or goat, nor meat of a dead or sick or leprous [animal], nor that of an animal that does not want to eat; no one shall sell mutton nor present castrated sheep, nor meat of a sow as that of a boar"), *Le Petit Thalamus de Montpellier* (ed. Pegat and Thomas 1840: 47).

256 "Cartulaire d'Albi" (ed. Vidal 1902: 462–463).

Fig. 7: Gathering chard, *Tacuinum sanitatis*, Bibliothèque nationale de France, nouvelle acquisition latine 1673, f. 27 recto. Source: gallica.bnf.fr.

> Ad home so necessarias ... quar sa carn li dona noyriment, sa pel et pels vestiments, so fems et urina a terra engrayshament, et breument no ha re en si que no sia convenient ad aysso o a medicament"[257]
>
> (They are necessary to man: for their flesh gives him nourishment, its skin clothing, its waste and urine nourish the earth, and, in short, there is nothing in them that is not useful or medicinal.)

Arnaldus de Villanova recommended goat meat for consumption throughout the year.[258] "Cabridet" appears only once in all of the *Concordance de l'occitan médiéval* 2005 and not at all in the *Trésor manuscrit de l'ancien occitan* 2017– as of 19 June 2022; it may be a hapax legomenon for this period; there is no question as to its meaning. Levy cites this line in our text and translates the term as "kleines Zicklein", using a diminutive and an adjective, "a small kid or goat".[259]

Cansonetas (l. 247). The "-eta" suffix is a diminutive, so a "cansoneta" is a little "canso", a little song, though Joseph Anglade once defined "cansoneta" as a synonym for "canso", song.[260] The canso, often defined as a love song, was a dominant genre of troubadour performers;[261] the cansoneta would be a more light-hearted or light-weight genre. Two medieval performers, Sail d'Escola[262] and Albertet,[263] are described as having composed cansonetas in the vidas, troubadour biographies.

Capons (l. 334). Occitan author Matfre Ermengaud considered capon, a castrated or neutered cockerel (rooster), to be the best meat to serve at a Christmas feast.[264] The *Elucidari* describes the bird's flesh in these terms: "Sas carns sobre totas carns de volatils so competens, et de melhor noyriment engendro sanc de major puritat"[265] ("Its flesh, above all flesh from birds, is appropriate and of best nourishment, engen-

---

257 *Elucidari* (ed. Hershon, Ricketts et al. 2018: 517).

258 Carré 2019: 180. Arnaldus de Villanova, *Regimen sanitatis ad regem Aragonum*, chapter 13: "De alimentis que sumunter ex animalibus" (ed. García-Ballester and McVaugh 1996: 452–455).

259 Levy 1894–1924: 1.181 s. v. cabridet.

260 It is intriguing that Anglade translated "cansoneta" as "chansonette", but glossed the term as "chanson". See Peire Raimon, "Poésies du troubadour Peire Raimon de Toulouse" (ed. Anglade 1919: 189), and Peire Raimon, "Poesies du troubadour Peire Raimon de Toulouse, suite" (ed. Anglade 1920: 302).

261 See Guilhem Molinier, "Cansos es dictats que d'Amors / principalmen o de lauzors" ("'Canso' is a kind of speech which talks mostly about love or praise"), Guilhem Molinier, *Las Flors del Gay Saber* (ed. Anglade 1926: ll. 3121–3122).

262 The vida for Sail explains that he was from Bergerac and that he became a joglar, a performer who "fez de bonas cansonetas" ("he created good 'cansonetas'"), *Biographies des troubadours* (ed. Boutière and Schutz 1973: 64).

263 Albertet came from the region of Gap. Whether it is he or his father Asar, also a performer, who is credited with composing cansonetas is unclear in Albertet's vida: "Albertez si fo de Gapenses, fils d'un joglar qe ac nom N'Asar, que fon trobaire e fez de bonas cansonetas" ("Albertet was from the Gap region, son of a joglar whose name was Sir Asar; he was a troubadour and composed good cansonetas") *Biographies des troubadours* (ed. Boutière and Schutz 1973: 508).

264 See Pfeffer 2006.

265 *Elucidari* (ed. Hershon, Ricketts et al. 2018: 306).

dering blood of great purity"). Arnaldus de Villanova recommended capons that were six to eight months old.[266]

Carbons (l. 333). Charcoal and embers were a readily available source of heat for cooking in the Middle Ages, offering a more controllable heat than that of an open flame. See "raust and raustir", below.

Carbonadas (l. 336). A "carbonada" is a meat dish, prepared over coals (whence the name), probably using a grill. Levy points to a quotation from the *Floretus* glossary which describes the cooking method:[267] "Carbonada frixa, pars carnis quae super carbones coquitur" ("fried carbonada, a piece of meat which is cooked over coals").[268] In this case, the -ada suffix points to a method of preparation rather than to an ingredient. The term also appears in a late-medieval Occitan mystery play.[269]

Cauls (ll. 322, 406). Cabbage was the foundation plant of every medieval Occitan garden.[270] The *Elucidari* describes it as dry and warm, available in summer and winter, with medicinal uses, though its author calls cabbage undigestible, "lor substancia es grossa et mal digestibla" with various undesirable side effects, including constipation.[271] These ideas are seen in lines 405–408, where *Occitan Health Advice* argues against eating too much cabbage. Arnaldus de Villanova advises consuming most vegetables at the beginning of the meal and calls cabbage "styptic".[272] The Montpellier medicinal recipe book lists cabbage among foods that are "constrictivas", useful as constrictors in treatment of pleurisy.[273] Cabbage was incorporated into any number of medicinal recipes, a fact that points to its availability.[274] Medieval illustrations of cabbage suggest that those gardens had more loose-head cabbage than the round cabbage we find in today's markets.

Clavels (l. 398). As Melitta Weiss Adamson notes, "In medieval Europe cloves were a highly sought-after spice appearing in countless upper-class preparations ranging from meat and fish dishes to fruit mousses and gingerbread".[275] Cloves appear in the *Modus*, in a recipe for a parsley sauce,

---

266 Carré 2019: 180; Arnaldus de Villanova, *Regimen sanitatis ad regem Aragonum*, chapter 13 "De alimentis que sumuntur ex animalibus" (ed. García-Ballester and McVaugh 1996: 453).

267 Levy 1894–1924: 1.211 s. v. carbonada.

268 "Vocabulaire provençal-latin" (ed. Blanc 1891: 57).

269 "Hy so meses en grans brasyers / la hont los rostiso a payroladas / los demonis coma carbonadas" ("Here are placed the large braziers where they roast demons in small cauldrons [as if they were] grilled meats") *Les Mystères provençaux* (ed. Jeanroy and Teulié 1893: ll. 2361–2363).

270 See Pfeffer 2016: 63–64 and 195, citing Stouff 1996: 74; see also Adamson 2004: 8.

271 *Elucidari* (ed. Hershon, Ricketts et al. 2018: 460).

272 Carré 2019: 179; Arnaldus de Villanova, *Regimen sanitatis ad regem Aragonum*, chapter 9, "De luguminibus" (ed. García-Ballester and McVaugh 1996: 440–441).

273 Brunel 1957: 308, recipe number 40.

274 See Corradino Bozzi 1997: 428–429 Glossario s. v. caul and caul rog. Also "A Fourteenth-century Provençal Receptary": recipe number lxxv and p. 222.

275 Adamson 2004: 19.

Fig. 8: Cabbage, *Tacuinum sanitatis*, Bibliothèque nationale de France, nouvelle acquisition latine 1673, f. 29 recto. Source: gallica.bnf.fr.

Ad faciendum salsam julverti simplicem vel compositam, recipe de julverto, de pane et unum parum de alliis. Et mole et distempera cum aceto. Si vis facere compositam, pone gyngibre, clavellos, canellam et unum parum de vino.[276]

(To make a simple or mixed parsley sauce, take parsley, bread, and a bit of garlic. Grind [these together] and dilute with vinegar. If you wish to make [it] a mixed sauce, add ginger, cloves, cinnamon and a bit of wine.)

The spice was also included in a number of pharmaceutical recipes, particularly in the Chantilly collection of texts.[277]

Cobertor (l. 341).[278] A late medieval Occitan manual for the abacus uses blankets in one of its math problems; one is three "canas" long and two "canas" wide, the other five "canas" by three "canas".[279] As one cana or canna was approximately two meters long,[280] these sample blankets were probably fictional, at 18 feet × 12 feet and 30 feet × 18 feet (9 m × 6 m; 10 m × 9 m); the examples from the abacus manual do tell us that blankets were rectangular rather than square. Manuscript illustrations suggest smaller blankets. Scholar Édouard Forestié (1847–1911) noted that some blankets were luxury items, sometimes left as a pledge on a loan.[281]

Colra negra (l. 378). Black choler, bile or melancholy, is one of the four temperaments associated with Galenic medicine (see above, Galenic Tradition and Humoral Theory). Matfre Ermengaud reports that choler is "born of fire", "colera es la premieira, / la quals nais del fuoc entieira."[282]

Cominada (l. 94). The spice "cumin" usually presents as "comin" or "comi". Emil Levy understood "cominada" as a drink made with cumin;[283] his only source is our text. The -ada suffix could also point to a dish or confection made with cumin.[284] The *Elucidari* approves of the spice, warm and dry to the second degree, and describes

276 Lambert 1989b: 137; recipe number 6 for Salsa julverti; *Modus*: f. 24 verso–25 recto.

277 We find in this manuscript a remarkably detailed recipe for "piment", spiced wine, which includes 20 cloves, along with weights for other spices, to be combined with white wine. See Corradino Bozzi 1997: 340.

278 Wartburg 1922–2002: s. v. COOPERTŌRIUM, 2: 1152.

279 Francés Pellos, *Compendion de l'abaco* (ed. Lafont 1967: 122).

280 Per Paden 2016: 182 s. v. corda, agreeing with Pansier 1924–1932: 3.34a, s. v. cana, canna.

281 "Ces couvertures, qui servaient à divers usages, soit en guise de housse pour les chevaux, soit pour les livres, soit pour couvrir les lits, étaient armoriées quand elles appartenaient à des nobles. Il semble que ces objets de luxe devaient avoir une certaine valeur, puisque la plupart des citations que nous relevons à ce sujet sont des nantissements: le prieur de Léojac, P.-G. de Penne, met en gage chez Bonis une couverture de laine avec ses armes (f° 2). Un damoiseau promet de donner un houcaran et une couverture de taffetas en gage (f° 64). Bertrand de Malhac dépose une couverture de taffetas armoriée de fleurs de lys (f° 65)," Bonis Brothers, *Livres de comptes* (ed. Forestié 1890–1894: lxxi).

282 Matfre Ermengaud, *Breviari 2* (ed. Ricketts 1989): ll. 7689–7690.

283 Levy 1894–1924: 1.298b s. v. cominada.

284 See Pfeffer 2022.

Fig. 9: A Checkered Blanket, *Roman de Jaufre*, Bibliothèque nationale de France, fonds français, 2164, f. 54 verso. Source: gallica.bnf.fr.

the positive medical uses of cumin in some detail.[285] The Avignon medical recipe collection uses cumin in only one recipe, from which I infer a high cost for the spice.[286] On the other hand, the Montpellier medical recipe collection incorporates ground cumin in seven different recipes;[287] the collections edited by Corradini Bozzi mention cumin at least seventeen times.[288]

Cogorletas (l. 286). Gourds and squash were well known in the Mediterranean world, though varieties like zucchini (*Cucurbita pepo*) are themselves species with New World origins, unknown in medieval Europe. Ibn Butlan suggested that gourds eased thirst and that the vegetable also calmed a cough and a fever.[289] Perhaps with

285 *Elucidari* (ed. Hershon, Ricketts et al. 2018: 433).
286 Brunel 1966: 512–513, recipe number 12.
287 Brunel 1957: 300, 303, 304, 305, 307, and 315, recipe numbers 3, 15b, 16, 20a, 24, 32b, and 102.
288 Corradini Bozzi 1997: 431, Glossario, s. v. comin.
289 Ibn Butlan, *Taqwim al-Sihha* (ed. Elkhadem 1990: 170–171).

this counsel in mind, one collection of Occitan medical recipes advises using squash seeds, "grana de la cogorlas", in the recipe for a syrup for pleurisy.[290] It is an interesting example of transmission that the Catalan scribe of V1 translated the rhyme word "cogorletas" into his language, writing "carabaçes".[291] The author of *Occitan Health Advice* may have coined the diminutive form simply for reasons of rhyme; the diminutive appears to be a medieval hapax legomenon.[292]

Descortz (l. 248). Though the first meanings of "descort" are "discord, dissonance",[293] the descort was also one of the several genres of lyric poetry defined in the *Leys d'amors*:

> Descortz es dictatz mot divers e pot haver aytantas coblas coma vers, so 's assaber de .v. a .x., las quals coblas devon esser singulars, dezacordablas e variables ... e deu tractar d'amors o de lauzors o er maniera de rancura.[294]
>
> (A descort is a lyric that can have as many stanzas as the *vers* genre, that is between five and ten, which stanzas should be unique, discordant, and variable. ... and it should treat of love or praise or grudges).[295]

Troubadour Garin d'Apchier is credited as having created the genre,[296] practiced by others as well. The genre is included in the Occitan romance *Flamenca* in the list of things a good entertainer should know.[297]

---

290 Meyer 1903: 280.

291 Institut d'estudis catalans, *Diccionari*: s. v. carabassa.

292 Our text provides the only example of this word in the *Concordance de l'occitan médiéval* 2005. The noun "cogorda" is found in the *Trésor manuscrit de l'ancien occitan* 2017–, but not the diminutive. Adams 1913: 190, lists "cogorleta" as a diminutive of "cogorda"; he adds that "cogorla" is cited by Mistral 1878 (Adams 1913: 190n4), though Adams appears not to have located Mistral's citation. Levy 1894–1924: 1.274a, s. v. cogorleta, points to Mistral's "coucoudeto, cougourleta", "petite courge". Wartburg 1922–2002: s. v. CŬCŬRBĪTA, 2: 1459b, may be implicitly pointing to *Occitan Health Advice* in its listing of "cogorleta" "anf. 13 jh."

293 Raynouard 1836–1845: 2.483b, s. v. descort.

294 Guilhem Molinier, *Las Leys d'amors* (ed. Fedi 2019: 374).

295 The verse *Leys d'amors* use similar language: "Descorts es dictats mot divers / d'aytantas coblas com ha vers" ("The descort is a very diverse lyric, of as many stanzas as there are lines"), Guilhem Molinier, *Las Flors del Gay Saber* (ed. Anglade 1926: ll. 3165–3166).

296 "Garins d'Apchier si fo uns gentils castellans de Jauvadan, del evesquat de Meinde, qu'es en la marqua d'Alverne e de Roserge e del evesquat del Puoi Sainta Maria. Valenz fo e bons guerrers, e larcs e bons trobaire e bels cavalliers; e saup d'amor e de dompnei, e tot so qu'en era. E fetz lo premier descort que anc fos faiz." ("Garin d'Apcher was a gentle nobleman of the Gévaudan, of the bishophric of Mende, which is on the border of Auvergne, Rouergue and the bishophric of Puy Sainte-Marie. He was valiant and a good warrior, generous and a good composer and a fine knight. He knew of love and gallantry and all that related to these. And he composed the first descort ever."), *Biographies des troubadours* (ed. Boutière and Schutz 1973: 343).

297 "Qui saup novella violadura, / ni canzo ni descort ni lais, / al plus que poc avan si trais" ("He who knew a new tune on the vielle, a lovesong, a descort, or a lai, he tried to put himself ahead"), *Flamenca* (ed. Gschwind 1976: ll. 596–598).

Diantes (l. 96). Édouard Forestié understood "diantos", similar to "diantes", as "carnation extract".[298] Emil Levy and the *Dictionnaire de l'occitan médiéval* (s. v. diantes) connect the term to maidenhair fern (*Adiantum capillus-veneris*).[299] A plant material, ground fine, could be added to a syrup or other liquid.[300] The word is not found in the Montpellier medical recipe book nor in the Avignon collection; "diantos" is used by Matfre Ermengaud in his discussion of plants (ll. 6955–6956).[301]

Dyarrodon (l. 92). This pre-modern pharmaceutical product made with red roses is described as a fourteenth-century Occitan hapax legomenon in Wartburg 1922–2002.[302] Emil Levy defined the word as "Rosenlatwerge, Rosenpulver", rose powder.[303] Wartburg 1922–2002 references a medieval Occitan medical recipe collection which mentions "dirodon Julii",[304] slightly different from the reference in *Occitan Health Advice*. These two medieval citations demonstrate, however, that the word is not a hapax. The concoction was also described by Arnaldus de Villanova in the chapter on electuraries in his *Antidotarium*.[305]

Empebrat (l. 338). The word clearly means "seasoned with pepper"; Mistral 1878 translates the infinitive "empebra" as "to pepper" (parallel to "to salt"). It is a unicum in the *Concordance de l'occitan médiéval* 2005 and marked as a hapax legomenon in Wartburg 1922–2002.[306] The prefix "em-" is common in Occitan; its attachment to "pebrar" is no surprise. Adams suggests that there was not always "a tangible distinction between simple word and compound"[307] as is the case here. It is possible that *Occitan Health Advice*'s author created the word for reasons of meter.

Escatz (l. 242). The game of chess came to Western Europe via several routes, appearing in the West well before the return of Crusaders from Palestine. The earliest Occidental reference is in a tenth-century Latin text, *Versus de scachis*, which describes the game as played at the time.[308] Chess is also described in the thirteenth-

---

298 Bonis Brothers, *Livres de comptes* (ed. Forestié 1890–1894: cxxvi).

299 Levy 1894–1924: 2.233b, s. v. diantos.

300 Consider Matfre Ermengaud's thoughts on the peony: "e dona.l om en beuratge / la pols ab suc de plantage" ("one gives it as a drink with juice from a plantain [*Plantago lanceolata*]"), Matfre Ermengaud, *Breviari 2* (ed. Ricketts 1989: lines 7067–7068).

301 See Ricketts 1989: 262.

302 Wartburg 1922–2002: s. v. RHODON, 10: 381b. Guiot de Provins, a late twelfth-century French author, includes "dyarodo julii" in his diatribe against doctors. See Guiot de Provins, *La Bible* (ed. Orr 1915: 91): l. 2619.

303 Levy 1894–1924: 2.233b, s. v. diarrodon.

304 Brunel 1957: 302 and 322, recipe number 11.

305 Arnaldus de Villanova. *Antidotarium* (ed. Taurelli 1585: XIII, column 413): s. v. Diarhodon abbatis and Diarhodon magnum, both made with roses and prescribed for very different ailments.

306 Wartburg 1922–2002: s. v. PĬPĔR, 8: 553b.

307 Adams 191: 432.

308 This poem exists in two manuscripts, both dated to circa 1000, housed in the Stiftsbibliothek of the Benedictine Abbey in Einsiedeln, manuscripts 319 and 365. See Gamer 1954: 740–744.

Fig. 10: Game of chess, *Libro de los juegos*, chess problem #17, Patrimonio Nacional, Real Biblioteca del Monasterio de El Escorial, RBMECAT T-I-6, f. 16 verso. With permission of the library.

century Spanish *Libro de los juegos*.[309] The rules of chess have changed over time, as the ability of pieces to move has increased (pawns gained the ability to advance two squares on their first move; the queen replaced the bishop and became the most powerful piece).[310] The *Libro de los juegos* portrays men and women playing the game and includes images that suggest parents teaching children how to play. Jean-Michel Mehl documents chess as a game whose prime medieval players were members of the nobility or upper bourgeoisie.[311]

Espic, erba del (l. 410). While "espic" can mean "ear", the grain-bearing tip or spike on the stem of a cereal plant, the text refers to an herb in the lavender family.[312] One species of the lavender genus, *Lavandula latifola*, is known as spike lavender; its scent is stronger and more pungent than that of *Lavandula angustifolia*, true lavender. The Montpellier medical recipe collection uses multiple spices including "espic",

309 The oldest manuscript copy of this work is housed in the monastery of San Lorenzo de El Escorial, Real Biblioteca del Monasterio, T.I.6.

310 See Mehl 2010: 137–145, for a discussion of women as pieces and players of the game of chess.

311 Mehl 2010: 289–290.

312 See Wartburg 1922–2002: s. v. LAVARE, 5: 219 and s. v. SPĪCA, 12: 174.

glossed by Brunel as lavender, in its recipe for "gingibrat".[313] The same spice is used in four different concoctions in the Avignon pharmaceutical collection.[314]

Fenholh (l. 366). Fennel is used frequently in medieval medications;[315] medical recipes from Montpellier call for fennel seed, root, or juice in eleven different preparations.[316] Arnaldus de Villanova discusses a "fennel wine", probably wine infused with fennel.[317] This plant is not part of the Bonis pharmacopoeia, perhaps because it was grown locally and therefore so inexpensive as not to warrant mention in their records. It is also not mentioned by Ibn Butlan, though it may have been added to later versions of his text. The *Elucidari* describes it as a medicinal plant, making clear that it exists in domestic gardens and in the wild, "Totas aquestas virtutz ha fenolh ortenc, mas le salvagge es de major eficacia"[318] ("Garden fennel has all these strengths, but wild fennel is more efficacious"). Matfre Ermengaud reports that fennel is good for the eyes, "fenoilhs coforta la vista" (l. 6959). The Ashburnham recipe collection attributed to Peyre de Serras uses fennel in six different concoctions for ocular problems.[319]

Figuas (l. 316). Figs are a "recommended food" for Ibn Butlan, who considers them moderately warm and moist.[320] The tree grows well in the Mediterranean area, and figs were part of the medieval Occitan diet.[321] Noël Coulet reported a 1430 Aix-en-Provence inventory that recorded two different types of figs, the "mussenqua" ("mussega") or "mouissone", and the "cotinhaqua".[322] It is telling that the *Oxford English Dictionary* reports that the English word "fig" < Old French "fige, figue", derived from the Occitan "figa, figua".[323]

The Bonis brothers in Montauban did a regular business in the sale of figs,[324] and the city government developed legislation regarding the locally-grown fruit.[325] A reference to figs in the romance of *Jaufre* suggests that they were not an expensive

313 Brunel 1957: 309, recipe number 42. See also "A Fourteenth-century Provençal Receptary", recipe number lx and p. 246.

314 Brunel 1966: 510, 513–514, 517, 521–523, recipe numbers 4, 15, 30, and 61.

315 See Brunel 1966: 514, 519, 521–522, recipe numbers 16, 17, 37, 53, and 57. Also Adamson 2004: 13.

316 Brunel 1957 calls for fennel seed, root, or juice in recipe numbers 2a, 3, 4, 16, 31a, 41c, 42, 48, 51, 69, and 73. Fennel is also used with frequency in the collections edited by Corradini Bozzi 1997: 442–443 *Glossario*, s. v. fenoil, fenoill.

317 Arnaldus de Villanova, *Le Livre des vins* (transl. Gifreu 2011): 102–103. Fennel wine is recommended in the Latin *Letter to Maecenas*; see Helmreich 1889: 12.

318 *Elucidari* (ed. Hershon, Ricketts et al. 2018: 444).

319 "A Fourteenth-century Provençal Receptary" (ed. Norwood 1978: 195–198 and 227, s. v. fenolh); recipe numbers xlj.2, xlj.4, xlj. 9, xlj.14, xlj.24, and xlj.27.

320 Ibn Butlan, *Taqwim al-Sihha* (ed. Elkhadem 1990: 150).

321 Ruas et al. 2005–2006: 201. Also, Adamson 2004: 24.

322 Coulet 1988: 125, who thought that this last fruit was "la figue rose de Cotignac".

323 *Oxford English Dictionary*, s. v. fig n.[1].

324 Pfeffer 2016: 76–77.

325 Forestié 1884: 14.

Fig. 11: Figs, *Tacuinum sanitatis*, Bibliothèque nationale de France, nouvelle acquisition latine 1673, f. 2 verso. Source: gallica.bnf.fr.

product, as the hero exclaims that his potentially dangerous behavior does not concern him, "non presa una figa" ("it is not worth a fig"), using the fruit in a proverbial expression for items of little value.[326] The *Elucidari*'s article on figs does not suggest that the author was personally familiar with the fruit.[327]

Gails (l. 334). Ibn Butlan suggests that roosters are best consumed during the winter,[328] advice that *Occitan Health Advice for the Layperson* follows. As a foodstuff deemed moderately warm and dry, the dish would serve to compensate for winter's cold and damp. The *Dictionnaire de l'occitan médiéval*, s. v. gal, does not give this spelling. Raynouard adds the "galh" and "gau" orthographies;[329] "gail" is equivalent to "galh". "Jal", found in Levy, means rooster's crowing rather than the bird itself.[330] The *Elucidari* devotes a paragraph to roosters, describing their behavior more than their culinary utility; the work does admit that the bird has some medical uses: "contra melancolica humor et flecmatica, contra grossa ventozitat d'estomach, contra inflacio et dolor d'arterias et contra longa febre erratica"[331] ("against melancholic and phlegmatic humors, against stomach winds [belching/burping], against swelling and pain in the arteries and against erratic and long fevers").

Gallinas (l. 334). Chickens and hens of all ages were, along with roosters, recommended by Ibn Butlan as a foodstuff for the winter months because of their humoral warmth.[332] Chicken was an affordable barnyard animal whose meat and eggs were available to many households in the Middle Ages; Catalan philologist and historian Antònia Carré (1960–) states that chicken was the meat most recommended by medieval doctors.[333] Arnaldus de Villanova recommends hens that have begun to lay eggs as the best, recommended for the autumn months.[334] The *Modus* uses chicken in seven different recipes.[335] The Montpellier medicinal recipe book lists hens, "galinath", and old chickens, "polh vielh", among foods that are "constrictivas", useful as constrictors in treatment of pleurisy.[336]

Gigimbre (ll. 320, 398, 412). Ginger was used as a spice and as an ingredient in medications in the Middle Ages; it is the spice which appears most often in the

---

326 Pfeffer 2016: 271; *Jaufre* (ed. Lee 2006: l. 4352).

327 *Elucidari* (ed. Hershon, Ricketts et al. 2018: 440).

328 Ibn Butlan, *Taqwim al-Sihha* (ed. Elkhadem 1990: 180).

329 Raynouard 1836–1845: 3.418a, s. v. gal, 3:418a, and 3.441b, s. v. gau 3:441b.

330 Levy 1894–1924: 4.17b, s. v. gal.

331 *Elucidari* (ed. Hershon, Ricketts et al. 2018: 306).

332 Ibn Butlan, *Taqwim al-Sihha* (ed. Elkhadem 1990: 180). Raynouard 1836–1845: 3.418a, s. v. galina, galinha.

333 Carré 2019: 180.

334 Carré 2019: 180. Arnaldus de Villanova, *Regimen sanitatis ad regem Aragonum*, chapter 13 "De alimentis que sumuntur ex animalibus" (ed. García-Ballester and McVaugh 1996: 452).

335 Lambert 1989b: 141, 149–154; *Modus* recipe numbers 18, 37, 39, 40, 41, 42, 44, and 45. See also Adamson 2004: 34.

336 Brunel 1957: 308. recipe number 40.

Fig. 12: Poultry, *Tacuinum sanitatis*, Bibliothèque nationale de France, nouvelle acquisition latine 1673, f. 68 verso. Source: gallica.bnf.fr.

account books of the Bonis brothers of Montauban,[337] and it appears in a sixteen different recipes in the *Modus*.[338] Documentation suggests that there were at least two qualities of ginger available in the late Middle Ages, white, "gingibre blan", and what may have been a product of lesser quality, used more for cooking, "gingibre de cozina".[339] The *Elucidari* describes it as warm and dry, wild and domesticated, adding that "le may blanc et plus fresc es meilhor et may agut" ("the whitest and freshest is better and stronger"); the encyclopedia recommends tempering it with pepper.[340] Ginger appears twice in the Avignon medical recipe collection[341] and in four recipes in the Montpellier collection.[342] Philologist Frances Norwood adds that the spice was "subject to adulteration and substitution".[343]

Gingibrat (l. 95). The term is defined as candied ginger. The Montpellier medical recipe collection offers a recipe for this sweet, combining soaked ginger root with honey and with cooked field eryngo root (*Eryngium campestre*), along with multiple other spices.[344]

Glop (l. 98). A glop is an imprecise measure of liquid quantity;[345] according to Mistral 1878 (s. v. gloup, glout), a "glop" can range in volume from a drop to a sip; the *Donatz proensals* defines "globs" as "a mouthful of some liquid".[346] The term appears

337 Pfeffer 2016: 108

338 Lambert 1989b: 135, 137–139, 141, 143–147, 149–155, 157; *Modus* recipe numbers 1, 6, 10, 12, 18, 22, 25, 28, 30, 31, 36, 38, 40, 43, 46, and 51.

339 Pfeffer 2016: 187n261, with reference to a 1408 document reporting a shopping trip to Montpellier in preparation for Lent: "Item tramesi Mondo Delafon a Monpelier per quere las causas jotz-scrichas per la provisio de carema a XVII de febrier: Portet: I cabas anguila corren, C polguars, I cabas per portar l'anguiala, IIas quartayrolas figuas, I cofet rayms, avelana XV lr. ris XL lr. pastel II gr., cezes blanx Ia carta, cezes brus Ia carta, cera vermelha za lr. sucre pa Ia lr. canela fina Ia lr. giroffle za lr. gualenguar za lr. gingibre blan Ia lr., gingibre de cozina IIas lr., estriat (?), papier V mas, dregieya comuna IIas lr. II balajos de rauba, amido IIII lr. za liassa guaranhs per estacar, I plen oyre d'oli, III sest. emina de sal, XX s. per son despens, foras la sivada. Costet tot ... XVIII lr. X s." ("Item, in preparation for Lent, on February 17, I sent Mondo Delafon to Montpellier to get the things written below: He carried a container of live eels, 100 [different] eels, a container to carry the eels, 2 measures of figs, 1 box of grapes, 15 pounds of hazelnuts, 40 pounds of rice, 2 grains of dye, 1 carton of white wax, 1 carton of brown wax, 1 pound of red wax, 1 pound of loaf sugar, 1 pound each of fine cinnamon, nutmeg, galengal, white ginger; 2 pounds of ginger for cooking, 5 reams of paper, 2 pounds of common dragées, 2 trims for robes, 4 pounds of flour, a package of nails for attaching, 1 full sack of oil, 3 measures of salt, 20 sous for his expenses except for rye [for the animals]. The whole cost 18 livres, 10 sous."); text found in Bedel 1996: 43.

340 *Elucidari* (ed. Hershon, Ricketts et al. 2018: 489).

341 Brunel 1966: 517 and 519, recipe numbers 30 and 39.

342 Brunel 1957: 302, 307, 309, recipe numbers 11, 34b, 34f, and 42.

343 "A Fourteen-century Provençal Receptary" (ed. Norwood 1978: 251).

344 Brunel 1957: 309, recipe number 42.

345 Wartburg 1922–2002: s. v. GLŬTTĪRE, 4: 17.

346 In the *Donatz proensals*, "globs" is used as an example of a rhyme in "-ops". The Occitan text glosses the word as "plenum os alicuius liquoris" ("a mouth full of some liquid"), Uc Faidit, *Donatz proensals* (ed. Marshall 1969: 221 and 403).

twice in the Princeton manuscript collection of recipes, each time describing a quantity of liquid, "fai beure un glop bo(n) en dejun"[347] and "beves (un) glob bon"[348] ("have [him] drink a good glug on an empty stomach" and "drink a good glug"). The word is used in modern Catalan with this meaning.[349]

Juglaressas (l. 246). There has been a fair amount of scholarly debate about female musical performers in the Middle Ages; that this text offers this technical term for these women confirms their reality, particularly as here the women are associated with music. The term is also found in documents from Montauban[350] and Montpellier.[351]

Jussel (ll. 319, 397). A "jussel" is defined as a sauce or broth; Wartburg 1922–2002 calls the term in our text a hapax legomenon,[352] though the word is not unique to this work, as it is also found in the *Modus*, recipe number 40.[353]

Kalendes (ll. 279, 280, 301). *Occitan Health Advice for the Layperson* uses one element of the classical Roman calendar system, where each month's first day was named the calends. The Romans counted up to each monthly marker: our April 22

---

347 Corradini Bozzi 1997: 165–166.

348 Corradini Bozzi 1997: 181.

349 See the *Diccionari català-valencià-balear*, s. v. glop, and the *Gran Diccionari de la llengua catalana*, s. v. glop[1].

350 Montauban's sumptuary laws in the *Livre rouge des consuls* offer a number of proscriptions that mark joglaressas and prostitutes as distinct from the other women of the community. For example, we read in a 1291 edict: "Que neguna dona d'esta vila ni de la honor no auze portar, … negu rosset en deguna de sas raubas, mas tant solamen de un palm de drap outra terra, e non plus; e que la rauba sia tota redonda, so es a saber que sia tant longua devan coma derreire, a bona fe, exceptadas jotglaressa e putas" ("That no woman of honor of this city dare wear … any red fabric on her dress, except narrower than the width of a hand, and no more, with the dress as long in front as in back, in good faith, … joglaressas and prostitutes excepted."), *Livre Rouge des consuls de Montauban*: f. 59 verso; Le Bret 1841: 412–413, cited by Levy 1894–1924: 4.260a, s. v. joglaresa. I thank Sarah-Grace Heller for her help with the Montalbanais materials.

351 The consuls of Montpellier included several mentions of joglaressas in their proscriptions regarding wedding celebrations, in 1227, for example: "Encaras establem que negun joglar ni neguna joglaressa non auze anar a novias de jorns ni de nuegz, ni auze anar en maison que haia dona jazen, mais trompas puescon anar a las novias" ("Furthermore, we establish that no male nor female performer dare go to wedding celebrations, neither during the day nor at night, nor go to the house where the lady lies, but trumpets can go to the celebrations."), *Coutumes anciennes de Montpellier* (ed. Widmayer 2004: 113). And further, "Encaras establiem que joglars ni joglaressas ni trompas non anon a novias de jorns ni de nuegz" ("Furthermore, we established that neither male nor female performers nor trumpets can go to wedding celebrations neither by day nor by night."), *Coutumes anciennes de Montpellier* (ed. Widmayer 2004: 114). These appear to be the references cited by Raynouard 1836–1845: 3.585b, s. v. joglaressa. The legislation was repeated several years later: "Encaras establem que jotglar, ni jotglars, ni jotglaressas, ni trompas non anon a novias de jorns ni de nuegs" ("Furthermore, we establish that no performer, no joglar nor joglaress [i.e. neither male nor female], nor trumpet shall go to wedding celebrations during the day or the night."). The document carries the date of 1253; see *Le Petit Thalamus de Montpellier* (ed. Pegat and Thomas 1840: 147).

352 Wartburg 1922–2002: s. v. JŪSCĔLLUM, 5: 85a.

353 Lambert 1989b: 151; *Modus* recipe number 30, f. 27 verso.

would be called the tenth day before the calends of May (the 22nd and the first are included in the count), whereas our text seems to follow a medieval practice of counting from the marker.[354] The term "calendor" "Christmas" is somewhat rare and points to the region from which the author came (see above, Language of the Scribe, regionally specific vocabulary).

Laig de vaca (l. 392), **laig de cabra** (l. 277). The invitation to drink cow's milk is interesting, as milk was not a regular part of the adult diet in the Middle Ages.[355] Ibn Butlan does discuss milk, suggesting that sheep's milk may be the best;[356] Arnaldus de Villanova would agree in this evaluation.[357] In the Middle Ages, most milk was converted into cheese, a way of preserving the product.[358] French historians Pierre Riché and Danièle Alexandre-Bidon suggest that for medieval consumers, cow's milk was considered hard to digest and would therefore be ill-advised.[359] The *Elucidari*, on the other hand, promotes cow's milk as "unctuoza et may nutritiva que pecorina" ("unctuous and more nutritious than that of goats"),[360] adding "layt de vaca, sobregrossa, … es mot nutritiva de stomach et de membres cofortativa, de calor de stomach et de fegge refregerativa et sana nafras de budels et de mayritz" ("cow's milk, very fatty … is very nutritious for the stomach and comforts the limbs, cools the heat of the stomach and the liver and heals intestinal wounds and genital problems").[361] The Montpellier medicinal recipe book lists heated milk, "lait cuecha", among foods that are "constrictivas", useful as constrictors in treatment of pleurisy.[362] Arnaldus de Villanova is fearful of milk products and limited their consumption to the period from April 15 to the end of May.[363]

Laychuguetas, lachuguas (ll. 278, 368). While we do not often think of lettuce as part of the late medieval diet, it most certainly was.[364] Noteworthy is the use of a diminutive in *Occitan Health Advice for the Layperson*, "little lettuces" or "young or baby lettuce". Lettuce is named more than once, perhaps because Ibn Butlan describes it as the most nutritious of all vegetables.[365] The *Elucidari* promotes lettuce

354 See Cappelli 1998: 25.
355 Adamson 2004: 45.
356 Ibn Butlan, *Taqwim al-Sihha* (ed. Elkhadem 1990: 172).
357 Carré 2019: 182; Arnaldus de Villanova, *Regimen sanitatis ad regem Aragonum*, chapter 14, "De alimentis que sumuntur ex humiditatis animalium" (ed. García-Ballester and McVaugh 1996: 456).
358 Pfeffer 2016: 83.
359 Riché and Alexandre-Bidon 1994: 75–77.
360 *Elucidari* (ed. Hershon, Ricketts et al. 2018: 590).
361 *Elucidari* (ed. Hershon, Ricketts et al. 2018: 591).
362 Brunel 1957: 308, recipe number 40.
363 Carré 2019: 182; Arnaldus de Villanova, *Regimen sanitatis ad regem Aragonum*, chapter 14, "De alimentis que sumuntur ex humiditatibus animalium" (ed. García-Ballester and McVaugh 1996: 456), "a medio Aprilis usque ad finem Madii".
364 Adamson 2004: 8, suggests otherwise.
365 Ibn Butlan, *Taqwim al-Sihha* (ed. Elkhadem 1990: 164). Pucci Donati 2007: 126 and 156 notes that other medieval authorities had a different opinion of the vegetable.

Fig. 13: Milking sheep, *Tacuinum sanitatis*, Bibliothèque nationale de France, nouvelle acquisition latine 1673, f. 56 verso. Source: gallica.bnf.fr.

as something good to eat and good as a medicine, "sa substancia es subtil, majorment fresca, per que es leu digestibla" ("its substance is subtle, largely fresh, so that it is easily digested"); this work seems to identify several different types of lettuce-like plants, from bitter endive, "serralia", to a plant with "fuelhas redondas et breus" ("round short leaves"), which the work says some call "heratea".[366]

Lettuce is named in the *Coutumes de Montréal* (Aude)[367] and in a 1409 Manosque decree punishing those who steal lettuce and other vegetables from home gardens;[368] Raynouard s. v. lachuga, laytuga,[369] references the *Elucidari* (f. 80 and 212) and cites Daude de Prades who served lettuce to hunting birds.[370] "Lach" and "lachuga", "lettuce", are both included in one late Occitan-Latin glossary.[371] The northern French fourteenth-century *Mesnagier de Paris* mentions lettuce from Avignon,[372] and Louis Stouff points to multiple references to the plant in fourteenth- and fifteenth-century Provence.[373] The *Herbier d'Auch*, a pharmaceutical treatise, describes lettuce as cold and humid, "frega e humida", using the humoral terms established in the *Tacuinum sanitatis*.[374]

Lectuari (l. 87). Bondurand explains that this medication was made of various powders and extracts, to which were added sugar or honey syrups.[375] Electuaries were a common prescription in the Middle Ages, frequently recommended for the beginning of the day as in our text. The commercial records of the Bonis brothers, pharmacists in fourteenth-century Montauban, record four different kinds of electuaries: "normal" (their most frequently prepared), "restorative" ("confortatien"), "laxative" ("lectuari laxatio", found the least number of times in their records), and prepared with rose-water.[376] Most of their entries offer no details on the preparation of an electuary, with one exception. We learn that in 1346, they used "una tortagua e d'autras causas", a tortoise and other spices, to prepare a very expensive medica-

366 *Elucidari* (ed. Hershon, Ricketts et al. 2018: 451).

367 Cited by Levy 1894–1924: 4.297a, s. v. lachuga.

368 "Item, e per cascun caul, cogorda, concombre, lachuga et de tota autra ortolalha pagon, aysins quant dessus es dich, al senhor et al denunciant, e lo dan donat." ("Item, for each cabbage, gourd, cucumber, lettuce, and for any other garden vegetable they pay, as is said above, to the lord and to the informant, and the damage given.") Meyer 1909: 379.

369 Raynouard 1836–1845: 4.6b, s. v. lachuga, laytuga.

370 Raynouard 1836–1845: 4.6a, s. v. laxugeta, citing Daude de Prades, *Dels auzels cassadors* (ed. Schultz 1945): l. 2845.

371 Pansier 1924–1932: 5.111.

372 *Mesnagier de Paris* (ed. Brereton and Ferrier 1994: 420). The *Mesnagier*'s author describes the Avignon lettuce as better and more tender than that found in the Paris region.

373 Stouff 1996: 74–75 especially.

374 The *Herbier d'Auch*, in Corradini Bozzi 1997: 193. The *Tacuinum* describes lettuce as cold and humid, both to the third degree. It adds that the plant is useful to combat insomnia and gonorrhea and that it harms sexual vigor and sight. See Ibn Butlan, *Taqwim al-Sihha* (ed. Elkhadem 1990: 164).

375 Bondurand 1895: 195n.

376 See Bonis Brothers, *Livres de comptes* (ed. Forestié 1890–1894: cxxiii–cxxiv).

tion.[377] The Avignon recipe collection offers a recipe for "Electuarium de sucre rosat";[378] the recipe collection from the Pyrenees includes a detailed recipe for an electuary intended to heal *mal de cor*, which I understand as heartburn; this recipe includes cornflower and honey as ingredients.[379]

Legums (l. 322). Vegetables formed a larger part of the medieval diet than we generally think, because medieval documentation of vegetables can be harder to find – what grows in the adjacent garden is not noted in the written record. Fava beans, peas, and chickpeas were all cultivated in medieval Occitania, appreciated for their availability as a fresh product and as vegetables that could be dried and consumed in leaner months of the year.[380] Literary references to legumes are explicit in the *Breviari d'amor*[381] and in the romance of *Flamenca*.[382]

Lentilas (l. 356). Ibn Butlan reports that lentils are mildly cold and dry;[383] the *Tacuinum sanitatis* adds that lentils interact with the digestive system and, depending on their preparation, can lead to constipation or diarrhea.[384] The Montpellier recipe collection offers contradictory advice on the utility of lentils, listing them as foods that "constrict" and as foods that cause swelling.[385] Of the several texts edited by Corradini Bozzi, only those in the Chantilly manuscript use lentils much.[386] Their mention in *Occitan Health Advice* is noteworthy, as consumption of lentils seems to have declined in Occitania over the course of the Middle Ages.[387] The *Elucidari* describes the best lentils as those "que so may grossas, frescas et may cuechivas"[388] ("those that are fairly large, fresh, and well cooked").

Lugor (l. 60). The eytmon, LŪCOR, means "light" or "spark", the usual meaning of "lugor" in Occitan. However, this meaning does not make perfect sense in the context of our work. Carl Appel suggested "Glanz (der Augen u.s.w.)" ("luster [of the eyes

377 While electuaries were not generally very expensive, ranging from 6 to 12 sous, Bonis Brothers, *Livres de comptes* (ed. Forestié 1890–1894: cxxiii–cxxiv), this one cost 1 livre, 8 sous, Bonis Brothers, *Livres de comptes* (ed. Forestié 1890–1894: 121).
378 Brunel 1966: 512, recipe number 11; this collection also offers a recipe for an "Electuari de codon", a quince electuary, Brunel 1966: 525, recipe number 74.
379 Brunel 1956: 88–89.
380 See Adamson 2004: 8–9.
381 Matfre Ermengaud, *Breviari d'amor 4* (ed. Ricketts 2004: l. 17929).
382 *Flamenca* (ed. Gschwind 1976, l. 403).
383 Ibn Butlan, *Taqwim al-Sihha* (ed. Elkhadem 1990: 160).
384 Ibn Butlan, *Taqwim al-Sihha* (ed. Elkhadem 1990: 161).
385 Brunel 1957: 308, recipe number 39, "Aquestas son constrictivas" and 40, "Causas lazativas". "A Fourteenth-century Provençal Receptary" incorporates lentils in a recipe for dysentery. See "A Fourteenth-century Provençal Receptary" (ed. Norwood 1978: 186 and 232, s. v. lentilhas), recipe i.2.
386 Corradini Bozzi 1997: 452, Glossario, s. v. lentilha.
387 Archaeobotanist Marie-Pierre Ruas notes that lentils are found in 83 % of Gallo-Roman archeological sites in comparison to 30 % of medieval sites (Ruas 1992: 23).
388 *Elucidari* (ed. Hershon, Ricketts et al. 2018: 453).

Fig. 14: Lentils, *Tacuinum sanitatis*, Bibliothèque nationale de France, nouvelle acquisition latine 1673, f. 45 recto. Source: gallica.bnf.fr.

etc.]”),[389] which Emil Levy interpreted as “frisches Aussehen?” (“fresh appearance”) though the question mark shows that he clearly had his doubts.[390] Nonetheless, this meaning of “fresh visage, clear complexion” does seem to fit the context. Édouard Bondurand understood the word as meaning “sight”, “voir plus clair”.[391]

Most (l. 394). The Latin word MŬSTUM,[392] “grape juice that has not yet fermented”, changes meaning slightly over time. The *Elucidari* says that “must” is new wine, “vi novel es dit most”;[393] Wartburg 1922–2002 points to a 1358 Provençal mention of “vin most”,[394] translated as “vin non fermenté” by Pansier.[395] The distinction between grape juice and “unfermented wine” is unclear to me. The modern definition of this grape by-product is “freshly crushed grape juice containing the skins, seeds, and stems of the fruit”, the first step in the wine-making process.

The *Elucidari* reports that most wines are very good for the health, but that new wine is not:

> Begut, engendra gros fum de sompnis tribul[acio] inductiu, ajusta malas humors, fa brugir le ventre, infla les budels, car mot es ventos et spumos per forsa de calor.[396]
>
> (Drunk, it engenders a large smoke leading to sleep difficulties, it adjusts bad humors, makes the stomach growl and inflates the intestines, because it is very windy and frothy on account of its heat).

Given the encouraging advice in *Occitan Health Advice for the Layperson*, we must believe either that its author believed must was healthy or that this author did not understand “must” as meaning new wine.

Moutos (l. 317). Mutton was not unfamiliar to the medieval table; some scholars believe it was consumed less often than pork and could be expensive.[397] However, mutton was a meat eaten frequently in Catalonia, which showed a “preference for mutton … shared with other Mediterranean regions”.[398] Catalan scholar Ramón A. Banegas López argues that there was an absolute preference for mutton over other red meats in late medieval Catalonia.[399] Paleobiologist Yves Lignereux points to archeo-

389 Appel 1920: 269a, Glossary, s. v. lugor.
390 Levy 1894–1924: 4.442a, s. v. lugor.
391 Bondurand 1895: 194.
392 Wartburg 1922–2002: s. v. MŬSTUM, 6.3: 274a.
393 *Elucidari* (ed. Hershon, Ricketts et al. 2018: 486).
394 Wartburg 1922–2002: s. v. MŬSTUM, 6.3: 271.
395 Pansier 1924–1932: 2.56.
396 *Elucidari* (ed. Hershon, Ricketts et al. 2018: 486).
397 Adamson 2004: 31–33. Adamson was not familiar with the work of Banegas López, however. Banegas López 2010 offers an explicit comparison of meat consumption in a number of European venues and documents the importance of mutton in the south, generally.
398 Piñer 2022: 190, citing Banegas López 2017.
399 Banegas López 2017: §3. The data presented in Banegas López 2010: 71 shows mutton as representing 70 % of meat consumption in mid-fifteenth century Barcelona.

Fig. 15: Preparing Must, *Tacuinum sanitatis*, Bibliothèque nationale de France, nouvelle acquisition latine 1673, f. 76 recto. Source: gallica.bnf.fr.

zoological evidence from the château d'Usson (Ariège) that suggests that mutton and/or goat represented more than 50 % of meat bones found at this Occitan site.[400] The French scholar's comparative chart of pork, sheep and goats, and cattle bones found in eleven different archeological sites across the south of France makes clear that sheep, i.e. mutton, predominated in a good number of these locales,[401] seconding the documentation of Louis Stouff for medieval Provence.[402]

Occitan Health Advice's recommendation of two-year-old mutton is interesting in light of Ibn Butlan's suggestion of one-year-old mutton as the preferred product.[403] The *Elucidari* notes that the characteristics of the meat change as the animal ages, "quan so antix, defalh en calor" ("when it is old, it loses heat").[404] The *Tacuinum sanitatis* judged mutton as very balanced, which would make it a logical foodstuff in autumn, as suggested by *Occitan Health Advice*, though the *Tacuinum* advises eating the meat in springtime.[405]

Pan dur et mal saboros (l. 445). This mention of hard, I infer stale, bread is intriguing. Medieval Occitan cuisine was very familiar with toasted bread, used particularly to thicken sauces or dishes; implied in the cooking recipes is that the toast crumbs added rather than detracted from the flavor of the sauce.[406] Explicit in *Occitan Health Advice*'s description is that stale bread was not appreciated.

Panadas (ll. 335, 395). Raynouard translated the term as "panade [a kind of bread soup], sorte de mets".[407] The word is derived from PANIS,[408] pointing to something made with bread.[409] The *Dictionnaire de l'occitan médiéval* translates the term as "pâté, tourte";[410] we should think in terms of a pâté en croûte, usually prepared by professional bakers in the Middle Ages.[411] A document from Manosque makes clear that panadas were baked in an oven;[412] similar documentation is found in the Archives of Narbonne.[413] Panadas could be found on the noble table though it was part

---

400 Lignereux 2011: 37–38.
401 Lignereux 2011: 41.
402 Stouff 1969: 1431–1448 and 1970: 179–180.
403 Ibn Butlan, *Taqwim al-Sihha* (ed. Elkhadem 1990: 176).
404 *Elucidari* (ed. Hershon, Ricketts et al. 2018: 501).
405 Ibn Butlan, *Taqwim al-Sihha* (ed. Elkhadem 1990: 177).
406 Lambert 1989b: 135, 139–140, 145, 148; *Modus* recipe numbers 1, 2, 12, 13, 14, 15, 26, and 34.
407 Raynouard 1836–1845: 4.405b, s. v. 2. panada.
408 Wartburg 1922–2002: s. v. PANIS, 7. 546a.
409 See Pfeffer 2022.
410 *Dictionnaire de l'occitan médiéval*, s. v. panada.
411 See *Libre de Sent Soví* (ed. Grewe 2004: 20–21).
412 Cited by Levy 1894–1924: 6.42a, s. v. panada, "Que quant que panadas si farian e se coirian els forntz de Manoasca …" ("That when panadas are made, they are baked in the ovens of Manosque …").
413 Cited by Mouynès 1871–1879: 2.38–39, "… cascun fornier sia tengutz de coire quascuna setmana alcun servisi a quascuna maison, .ii. fromaggadas, o .ii. panadas de carn o de peis, e .ii. padenadas de carn o de peis, sian grandas o paucas" ("… every week, each baker must bake for each household, 2 fromaggadas [cheese dishes] or 2 panadas of meat or fish, and 2 skilletfuls of meat or fish, whether big or small").

Fig. 16: Butchering Sheep, *Tacuinum sanitatis*, Bibliothèque nationale de France, nouvelle acquisition latine 1673, f. 61 verso. Source: gallica.bnf.fr.

of medieval fast food as well.[414] I translate the term as "empanada", a turnover with a sweet or savory filling which is then baked or fried, whose Spanish name has moved into the American vernacular. The Catalan cookbook *Libre de Sent Soví* includes several recipes for "panades",[415] and the Occitan romance *Jaufre* describes a meal where partridge panadas formed part of the menu (see below, Culinary Context).[416]

Pebre (l. 398). Pepper was one of the most eagerly sought spices in the medieval period. Grown in south and southeast Asia, it was brought to Europe by land and by sea; the travel costs made it relatively expensive for consumers.[417] Nonetheless, as a prestige spice, it was much sought after, used in cooking and in medicinal preparations. The spice is included in no fewer than twenty of the *Modus*'s fifty-one recipes[418] and is found in medical recipe collections as well.[419]

Pebrada (l. 412). Logically, this word should refer to a dish made with a lot of pepper,[420] though the word can be an adjective describing the spicing of a particular dish as is implied in *Occitan Health Advice for the Layperson*. Occitan literary references suggest that "pebrada" may also refer to a kind of pepper sauce.[421]

Pigmens (l. 339). Bondurand describes piment as a drink made of honey and spices;[422] he omitted an important element from the beverage, wine. The *Elucidari* offers this description: "Pigment es dit quar si far de specias suptilment piladas, de las qual part pigmentaria si fan alcus beuragges et electuraris"[423] ("piment is called that because it is made of finely ground spices, from which spices are made various drinks and electuaries"), again omitting the wine. The registers of the Bonis brothers in Montauban offer several "recipes" for piment, employing various combinations of spices, notably ginger, cinnamon, cloves, grain of paradise, and honey; a piment concocted for the pope in Avignon used sugar instead of honey as the sweetener.[424]

---

414 Bruno Laurioux reminds us that many medieval households did not have the ability to prepare meals (for lack of space) and that the equivalent of take-out existed in medieval towns and cities, see Laurioux 2004 for a number of examples.

415 Santanach 2008: 136–139, recipe numbers XLVII and XLVIII; from the *Llibre d'aparaellar de menjar* (Santanach 2008: 196–197, 206–209), recipe numbers 37, 50, and 52.

416 *Jaufre* (ed. Lee 2006: l. 4238).

417 Adamson 2004: 16.

418 Lambert 1989; *Modus* recipe numbers 1, 2, 10, 12, 13, 15, 18, 19, 28, 31, 32, 34, 35, 36, 38, 42, 43, 44, 45, 46.

419 "A Fourteenth-century Provençal Receptary" (ed. Norwood 1978: 239).

420 See Pfeffer 2022.

421 See Ramon Vidal de Besalú, "Aital salsa, aital pebrada / sabetz vos far als non-gardans" ("Such a sauce, such a pepper dish, you know how to make for the unsuspecting"), *Judici d'Amor* (ed. Field 1991: ll. 884–885) and the *Roman d'Esther*, "aqels vengron a plens pairols; / bueu e mouton venc am pebrada, / amb eruga e am mostarda, ("Then came filled platters of beef and mutton with a pepper [sauce], with a rocket [sauce] and with mustard"), *Roman d'Esther* (ed. and transl. Thiolier and Notz-Grob 1997: ll. 134–136).

422 Bondurand 1895: 202n1.

423 *Elucidari* (ed. Hershon, Ricketts et al. 2018: 588).

424 See Pfeffer 2016: 316–317.

Arnaldus de Villanova offers a recipe for "nectar" which is, in fact, a piment:

> Prenez deux drachmes de gingembre choisi, girofle, cinnamome, et une drachme de poivre de Guinée. Mettez dans un demi-litre de bon vin, un vin grec, le meilleur. Ajoutez du miel ou du sucre dans un grain du mélange. C'est excellent.[425]
>
> (Take two drams of select ginger, cloves, cinnamon, and one dram of Guinea pepper. Put in a half-liter of good wine, Greek wine is best. Add honey or sugar to a grain of the mixture. It is excellent.)

As Arnaldus's inclusion of a recipe makes clear, piment could be used for medicinal purposes as well as for entertaining; the Bonis registers include a prescription for piment written by physician W. de Rabastens for his patient, a Mister Pons.[426] The Chantilly *Thesaur de pauvres* offers a recipe for piment remarkable for its detail.[427]

Polieg (l. 381). Pennyroyal is a plant in the mint family used frequently in medieval cooking and also in pharmacy.[428] The *Elucidari* describes it as "mot aromatica" ("highly aromatic"), used to cure a variety of ailments.[429] Curiously, the orthography of our text does not appear in the *Dictionnaire de l'occitan médiéval*, s. v. poleg, puleg, pulegi, though that resource elsewhere cites the work of Corradini Bozzi, who uses "polieg" as a headword in her glossary.[430] The recipe collection from the Pyrenees offers the spelling of "bolieg",[431] also not included in the *Dictionnaire de l'occitan médiéval*. Levy 1894–1924 cites *Occitan Health Advice* but does not use this spelling as a headword.[432] The orthography is included in Wartburg 1922–2002: s. v. PŪLĒIUM, 9: 521a.

Pollas (l. 318), **pols** (l. 276). As noted above, s. v. gallinas (l. 334), chicken is another meat considered balanced by the *Tacuinum sanitatis*[433] and therefore appropriate for consumption in a temperate season such as autumn.

Pulment (l. 174). Wartburg 1922–2002 defines "pulment" as an "extra" ("zukost"), a sauce, relish or condiment eaten with bread.[434] Curiously, this dictionary identifies the word in Old and Middle French texts but cites no medieval Occitan sources. Emil Levy points to the word in the *Chanson d'Antioche* and in various secular docu-

---

425 Arnaldus de Villanova, *Le Livre des vins* (transl. Gifreu 2011: 135).

426 "Item deu per 1ª pimentas que fezem per Mº Po[ns], que ordenec Mᵉ W. de Rabastens, a VII d'abril [1346]" ("Item, we are owed for one piment that we made for Mr. Pons on the order of Master W. of Rabestens"), Bonis Brothers, *Livres de comptes* (ed. Forestié 1890–1894: 2.9).

427 See Corradino Bozzi 1997: 340.

428 Brunel 1966: 512–513, recipe number 12; "A Fourteenth-century Provençal Receptary" (ed. Norwood 1978: 190, 197, 198, and 240, s. v. pulegi); recipe numbers xviiii 2, xli 20, xli 27, and lxvi 1.

429 *Elucidari* (ed. Hershon, Ricketts et al. 2018: 467).

430 Corradini Bozzi 1997: 467, Glossario, s. v. polieg rial.

431 Brunel 1956: 88, in recipe 531, for bad breath.

432 Levy 1894–1924: 6.428b, s. v. poleg.

433 Ibn Butlan, *Taqwim al-Sihha* (ed. Elkhadem 1990: 180).

434 Wartburg 1922–2002: s. v. PŬLMĔNTUM, 9: 545.

ments.[435] "Pulmen, polmen" is found in multiple Occitan texts with the larger meaning of "foodstuff". In contrast, two Latin-Occitan glossaries define the word as "companion" or "company" rather than as something to eat.[436] A third glossary defines the word as "nectar" or "piment", a spiced wine;[437] I believe this last glossary has confused the words "pulmen" and "pimen".[438] I have not been able to find the Occitan sources implied but not cited by Wartburg 1922–2002.[439]

Raust and **raustir** (ll. 333, 335). The verb for roasting, "raustir", and the result, a roast, stem from a proposed Germanic etymon, *RAUSTJAN.[440] Roasting can be accomplished in a baker's oven or in a camp oven; Raimon de Perelhos describes roasting on a grill and on a spit in his *Voyage to St Patrick's Purgatory*, "los rostian sobre grans gresilhas de fer cremans (sus); los autres rostian en grans astes sobre lo fuoc"[441] ("[human souls] were roasting on large flaming-hot iron grills; the others were roasting on large spits over the fire"). The *Elucidari* notes that one should avoid fatty meats unless they have been roasted, "Carnas trop grassas hom no deu majar si no raustidas".[442] The *Modus* does not use this method of food preparation explicitly. Arnaldus de Villanova appears to recommend roasting or boiling as the preferred preparations for meat; he is opposed to frying.[443]

Razims (ll. 315, 394). Grapes were an integral part of the medieval Mediterranean diet;[444] their appearance in *Occitan Health Advice for the Layperson* comes as no surprise. The humoral qualities of warm and moist[445] make them logical parts of an autumn meal. *Occitan Health Advice*'s insistance on ripe or sweet grapes has no equivalent in the *Tacuinum sanitatis*, more interested in separating skin and seeds from the pulp.[446] The *Elucidari* describes the various colors of grapes: white grapes "noyrisho alcunament et so de bona digestio" ("nourish somewhat and are each to

435 Levy 1894–1924: 6.433b, s. v. polmen.

436 Pansier's "Glossaire latin B" is the same text as that in "Vocabulaire provençal-latin" (ed. Blanc 1891: 59).

437 Pansier 1924–1932: 5.131 s. v. pument.

438 Wartburg 1922–2002: s. v. PŬLMĔNTUM, 9: 545a, points the reader to Mistral 1878, perhaps s. v. pimen, pumen, pimo. See also Mistral 1878: s. v. pument which the author includes more as a possibility, "sorte de boisson, v. pimen".

439 The *Dictionnaire de l'occitan médiéval* also points to Katara 1966: s. v. pulment. Katara translates the term as "Brei".

440 Wartburg 1922–2002: s. v. *RAUSTJAN, 16: 682b.

441 Raimon de Perelhos, *Voyage au Purgatoire de st Patrice* (ed. Jeanroy and Vignaux 1903: 32); Levy 1894–1924: 7.52a, s. v. raustir.

442 This line from the *Elucidari* (ed. Hershon, Ricketts et al. 2018: 499) is cited in Raynouard 1836–1845: 5.49b, s. v. raustir.

443 Carré 2019: 180; Arnaldus de Villanova, *Regimen sanitatis ad regem Aragonum*, chapter 13 "De alimentis que sumuntur ex animalibus" (ed. García-Ballester and McVaugh 1996: 454), "[carnes] frixas vero omnino debent vitare" ("one should avoid fried meats").

444 See Adamson 2004: 22.

445 Ibn Butlan, *Taqwim al-Sihha* (ed. Elkhadem 1990: 150).

446 Ibn Butlan, *Taqwim al-Sihha* (ed. Elkhadem 1990: 151).

Fig. 17: Grapes, *Tacuinum sanitatis*, Bibliothèque nationale de France, nouvelle acquisition latine 1673, f. 2 recto. Source: gallica.bnf.fr.

digest"), whereas black grapes, perhaps meaning raisins, "so de dura digesto, mas plus coforto l'estomach"[447] ("are hard to digest, but comfort the stomach more"). The Montpellier medicinal recipe book lists grapes as a food that leads to swelling.[448]

Reubarbizat (l. 92). The root of this word is clearly "reubarbari", rhubarb. Turning the noun into an adjective is perhaps a mark of the *Occitan Health Advice* author's creativity; I do not think this form is found elsewhere in Occitan.[449] The Avignon collection of pharmaceutical recipes includes "reubabi" in one preparation and "rubabe" in another.[450]

Ruda (l. 358). Rue is described by the *Elucidari* as a very warm and dry herb with multiple medicinal uses, consumed as a beverage or cooked and eaten.[451] One Auch pharmaceutical recipe collection describes rue as "mot bona e cauda" ("very good and warm");[452] the same manuscript, Auch, Archives départementales du Gers I 4066, includes a concoction for rabid dog bites, "Pren de ruda e de sal e d'aylh, ben mals e fay ne polvera, e met sobre lo mal lo mati e·l ser, d'aq(ue)la polvera, e guera" ("Take rue and salt and garlic, grind well and make into a powder, and put this powder on the patient in the morning and evening, and he will be cured").[453] The Avignon collection of recipes uses rue, spelled *ruta*, in three different preparations;[454] the Montpellier collection uses rue in six different recipes.[455] Frances Norwood adds that rue is a narcotic and a stimulant.[456] Matfre Ermengaud reports several uses for rue, noting that "Ruda fai lo vis clar e bell / e fort coforta lo cervell" ("rue makes the face clear and pretty and comforts the brain")[457] and that "sucz de ruda" ("rue sap") placed in the nostrils maintains the blood.[458]

Safra (l. 320). Grown across medieval Occitania, we have documentation on saffron grown in Aix-en-Provence in 1430[459] and suggestions of saffron production in the Quercy region in the fifteenth century;[460] there also existed a vibrant trade in the

---

447 *Elucidari* (ed. Hershon, Ricketts et al. 2018: 484).

448 Brunel 1957: 308, recipe number 40.

449 Wartburg 1922–2002: s. v. REUBARBARUM, 10: 348.

450 Brunel 1966: 521–522, recipe numbers 55 and 61. Rhubarb is not found in "A Fourteenth-century Provençal Receptary" (ed. Norwood 1978).

451 *Elucidari* (ed. Hershon, Ricketts et al. 2018: 471).

452 Corradini Bozzi 1997: 196.

453 Corradini Bozzi 1997: 237. See also Corradini Bozzi 2023.

454 Brunel 1966: 512–513, 527, recipe numbers 12, 13, and 94.

455 Brunel 1957: 299, 305–306, 315, and 316, recipe numbers 2a, 24d, 27, 48, 104, and 116.

456 "A Fourteenth-century Provençal Receptary" (ed. Norwood 1978: 242, s. v. ruta). The plant is used in multiple recipes in Peyre de Serras's recipe collection found in the Ashburnham manuscript.

457 Matfre Ermengaud, *Breviari 2* (ed. Ricketts 1989: ll.7025–7026).

458 Matfre Ermengaud, *Breviari 2* (ed. Ricketts 1989: ll. 7033–7034).

459 Coulet 1988: 124.

460 Jean Lartigaut 2000: 318, found evidence of saffron culture in the Quercy region, particularly near rivers and streams. He suggests that Figeac documents identify the town as a commercial center for saffron, adding that saffron could be grown in the vicinity of that community or come to Figeac from other cities, such as Albi (*Le Quercy*, 318).

Fig. 18: Rue, *Tacuinum sanitatis*, Bibliothèque nationale de France, nouvelle acquisition latine 1673, f. 32 recto. Source: gallica.bnf.fr.

spice, grown in Catalonia and sold throughout southern France.[461] The description in the *Elucidari* suggests familiarity with the plant, "alcu safra es ortenc, autre oriental qui es plus noble" ("some saffron is from the garden; the other, from the east, is more noble"); the work describes the spice as warm and dry to the first degree, "mot val en medicinas" ("very useful in medicines").[462] Even as a local product, saffron was and remains an expensive ingredient.[463] The Avignon pharmaceutical collection incorporates saffron into five different recipes;[464] the Ashburnham collection uses it in four recipes;[465] the Montpellier collection uses it only once.[466] Carole Lambert considered saffron as a spice that served as a marker of medieval Occitan cuisine; the *Modus*'s recipes call for it at least thirteen times.[467]

Salpicat (l. 337). The verb "salpicar" stems from the noun "salt". Wartburg 1922–2002 initially defines "salpicar" as "éparpiller, répandre de la cendre" ("to scatter, to spread cinders"), citing a 1220 Rouergat source;[468] this meaning feels distant from the spice. The dictionary then lists meanings of "saupoudrer de sel, saler" ("to sprinkle with salt, to salt") from Occitan regions closer to the Mediterranean coast and its salt trade. Raynouard defines "salpicar" as "saupoudrer, éparpiller" ("to powder, to sprinkle") and points to Catalan, Spanish, and Portuguese cognates.[469]

Salvasina (l. 332). The *Dictionnaire de l'occitan médiéval* translates the term as "venison" but does not include this orthography, preferring "salvatgina, salvaizina, salvazina".[470] The term certainly includes animals such as hare and rabbit, as several of Levy's examples demonstrate.[471] The *Roman d'Esther* includes non-domesticated animals on its banquet menu, along with menu items similar to those of this text: "astas de capons e de galinas / e de totas autras salvazinas"[472] ("skewers of capons and hens and all other wild fowl").

Sagnar, sangar, and conjugated forms (ll. 351, 361, 364, 371). Bloodletting or phlebotomy was a standard medical procedure in the medieval period, tied directly to the theory of humors and humoral balance and "regularly practiced as a prophylactic

---

461 Pfeffer 2016: 15, pointing to Combes 1990: 109.
462 *Elucidari* (ed. Hershon, Ricketts et al. 2018: 434).
463 Pfeffer 2016: 108. See also Adamson 2004: 14.
464 Brunel 1966: 512, 513, 514, 523, 527, recipe numbers 9, 14, 17, 63, and 86.
465 "A Fourteenth-century Provençal Receptary" (ed. Norwood 1978: 189, 192, 195, and 242, s. v. saffra); recipe numbers xvii, xxviii.3, xl.2, and xli.2.
466 Brunel 1957: 304, recipe numbers 20a and 20b.
467 Lambert 2002: 74, and 1989b; *Modus* recipe numbers 10, 18, 28, 30, 31, 32, 34, 36, 38, 42, 45, 46, and 49.
468 Wartburg 1922–2002: s. v. SAL, 11: 78.
469 Raynouard 1836–1845: s. v. salpicar 5: 15=39a.
470 *Dictionnaire de l'occitan médiéval*, s. v. salvaizina.
471 Levy 1894–1924: 7.452b, s. v. salvatgina. Adamson 2004: 35, limits "venison" to meat from roe deer, fallow deer, and red deer.
472 *Roman d'Esther* (ed. and transl. Thiolier and Notz-Grob 1997: ll. 99–100).

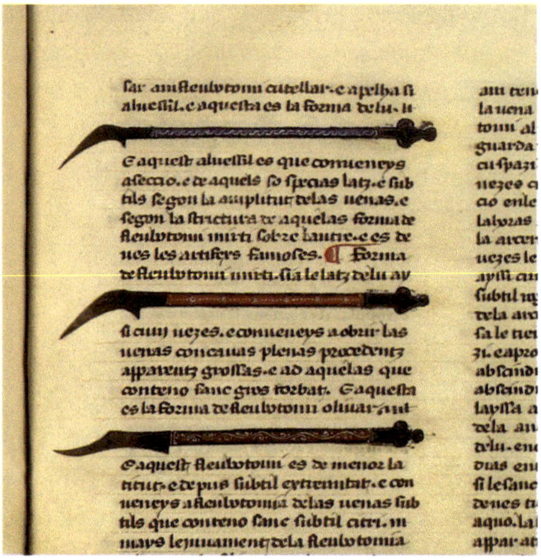

Fig. 19: Phlebotomy Tools, Albucasis, *Chirurgia*, Montpellier, Bibliothèque Universitaire Historique de Médecine, H 95, f. 52 recto. Université de Montpellier, Crédit photographique SCDI – Montpellier, Service photographique.

measure".[473] As Lanfrancus Mediolanensis (Lanfranc of Milan, circa 1250–1315) posited, "We use phlebotomy, generally, for three things: to preserve health, to protect from possible sickness, and to remove an existing illness".[474] The Occitan *Albucasis* devotes a long chapter 95 to phlebotomy (Montpellier, Bibliothèque Universitaire Historique de Médecine, Université de Montpellier, H 95, ff. 50 verso–54 recto), in which we learn of the many medical problems this treatment was deemed to cure.

Occitan Health Advice for the Layperson offers a fairly precise calendar of phlebotomy timing, specifying that January (l. 349), March (l. 351), May (l. 364), and summer (l. 289) are not good periods for bloodletting, and that April (l. 361), spring (l. 271), and November (l. 402) are recommended (see the discussion of *Regimina XII mensium*, above). Some of this timing is contradictory, as spring is considered a good time, but not in March or May. The Occitan *Albucasis* offers related advice, that bloodletting should be performed "al comensament de primaveyra, sobre dejun en primaviera, e autunpne e en invern aprop la comestio" ("at the beginning of spring, on an empty stomach in spring, and in autumn and winter after eating").[475] Montpellier physician Bernardus Gordonius (Bernard de Gordon) disagreed and suggested in his work on phlebotomy, part of his *Liber de conservacione*, that the concern with the calendar was "pure ignorance, sensationalism (curiositas), and malice".[476]

Scorrador (l. 352). The word as used here, meaning "cauter", a tool used in cauterization, appears to be somewhat rare. Ruggieri does not define it in her glossary.[477]

---

473 Wallis 2010: 281.

474 Translation by Michael McVaugh in Wallis 2010: 282, *Magna chirurgia*, tract 3.

475 Albucasis, *La Chirurgie d'Albucasis* (ed. Grimaud, rev. Lafont 1995: 198); *La Chirurgia* (ed. Elsheikh 1992: 201).

476 Demaitre 1980: 163.

477 Ruggieri 1930: 219.

Fig. 20: Scarification Tools, Albucasis, *Chirurgia*, Montpellier, Bibliothèque Universitaire Historique de Médecine, H 95, f. 3 recto. Université de Montpellier, Crédit photographique SCDI – Montpellier, Service photographique.

Levy 1894–1924 includes an entry for "escorredor",[478] a word whose meaning is unclear (he proposes "corridor" or "channel", neither of which seems appropriate here). The *Dictionnaire de l'occitan médiéval* includes "escorredor" (second entry) "inflammation", citing William Rothwell's reference to the Occitan prose *Chirurgia* (*Surgery*) of Rogerius of Salerno (ante 1140–circa 1195);[479] the English scholar connects this noun to the verb "escoriar". This translation does not work in *Occitan Health Advice*, where the verb governing "scorrador" is "posar" ("to lay" or "to place"). Jean-Pierre Chambon understands the word "escorredor" as found in the *Miracles que Dieus ha mostratz per sant Frances* to mean "cautère", "cauter", and seems to suggest that "scorredor" may not be connected to the verb "escorrer".[480]

The *Dictionnaire de l'occitan médiéval* links "escorredor" to *EXCORTICARE;[481] the meaning of cauter, a tool used for cauterization, seems clear and makes sense in context. This noun could be related to the verb "escorjar, escorgar, escorsar", "to skin, to scrape, to scratch".[482] As line 352 is from a Catalan witness, more evidence may be found in Catalan sources. Certainly, in the *Gran diccionari de la llengua catalana*, we find "escorxador", "ganivet per a escorxar", i.e. "paring knife" or similar utensil;[483] the parallels to cauter are evident. The reference in our text is clearly to some kind of medical tool, and cauter is a logical translation. Curiously, the Montpellier witness to the Occitan *Albucasis*, which devotes multiple folios to cauterization, does not use the term "scorrador"; it prefers the generic term "cauteri", several varieties of which are described and depicted.

Another possible explanation would be to connect "scorrador" to "scarificador", a tool used in a variation on bloodletting, involving scraping the skin to induce bleeding.[484] The Occitan *Albucasis* devotes a long chapter to cupping and connects this procedure to scarification, "Ventozacio es fayta segon doas manieyras: la una de lor es am scarificacio e extracio de sanc e l'autra es ventozacio ses scarificatio" ("Cupping is done in two ways. The first is with scarification and extracts blood; the second is cupping without scarification").[485] I would suggest that Catalan "escorxador", Provençal "scarificador", and Occitan "scorrador" are more closely related in meaning than has been recognized before.

---

478 Levy 1894–1924: 3.190a, s. v. escorredor.

479 Rothwell 1983: 497; Rogerius of Salerno, *Chirurgia*: f. 148 recto, b: "De escorredor de las borsas e de las emoroidas". See Binz 1908. The line cited by Rothwell does not appear in the Montpellier *Chirurgia* witness.

480 Chambon 1994: 571.

481 Wartburg 1922–2002: EXCORTICARE, 3: 282a.

482 *Dictionnaire de l'occitan médiéval*: s. v. escorjar. I thank Janet Gordon for this lead.

483 *Gran Diccionari de la llengua catalana*, s. v. escorxador2.

484 Pansier 1924–1932: 3.74, s. v. escarificar, escarificacion, "scarifier, scarification".

485 Albucasis, *La Chirurgie d'Albucasis* (ed. Grimaud, rev. Lafont 1995: 204); *La Chirurgia* (ed. Elsheikh 1992: 206).

Fig. 21: The Game of Tables, *Libro de los juegos*, Patrimonio Nacional, Real Biblioteca del Monasterio de El Escorial, RBMECAT T-I-6, f. 78 recto. With permission of the library.

Taulas (l. 242). There are multiple mentions of games of "taulas", tables, in the medieval Occitan literary corpus, starting with a song by the poet considered the first troubadour, Guilhem IX, "Ben vueill que sapchon li pluzor" (PC 183,2).[486] Described as "the favourite chamber game of the time",[487] "taulas" is a generic term for a game played with dice on a board which forms a sort of linear race track;[488] backgammon is one modern example of such games which were very popular in the medieval period. Alfonso X, king of Castile, Galicia, and León, commissioned the *Libro de los juegos*, completed in 1283, a book which describes rules for backgam-

---

486 See Levy 1894–1924: 8.76a-b, s. v. taula. Non-lyric examples include *Flamenca* (ed. Gschwind 1976: l. 302); *Daurel e Beton* (ed. Lee 1991: ll. 62, 1137, 1277 and 1566); Matfre Ermengaud, *Breviari 5* (ed. Ricketts 1976: l. 34025), among many; see also Murray 1941.

487 Murray 1941: 57.

488 Murray enumerates twenty-six different games as variations on the theme of tables, Murray 1941: 66–69. Note the difference between tables, played with dice, and dice games per se. See Mehl for details on dice games, which he describes as a condemned activity, despite their being "the prime entertainment of medieval Occidental society", "le divertissement majeur de la société médiévale occidentale", Mehl 2010: 179–180.

Fig. 22: The Game of Alquerque, *Libro de los juegos*, Patrimonio Nacional, Real Biblioteca del Monasterio de El Escorial, RBMECAT T-I-6, f. 92 recto. With permission of the library.

mon and other table games.[489] The game is often associated with chess in medieval Occitan literature.[490]

Taulier (l. 242). The game of checkers is frequently associated with that of chess in medieval Occitan literature (see above). A similar game, alquerque, is described in the *Libro de los juegos*. Today, checkers is played on the same playing board as that of chess.

Tractat (l. 1). It is possible that *Occitan Health Advice* provides the first use of this noun in Occitan; see Occitan Advice Literature and Genre Definition, below.

Ueos tenres (l. 276). Eggs were certainly part of the medieval Occitan diet within the confines imposed by fast days, observed somewhat less rigorously in southern France than in the north.[491] Remarkable here is the specification of "tender eggs", which is harder to interpret. Does the author mean "soft boiled eggs" or some other

489 See Alfonso X el Sabio, *Libro de los juegos* (ed. Orellana Calderón).

490 The romance of *Daurel e Beton* associates the two games repeatedly (ed. Lee 1991, ll. 1137, 1277, and 1566). The epic of *Rollan a Saragossa* also links the two games, *Roland occitan* (ed. and transl. Gouiran and Lafont 1991: l. 414).

491 Pfeffer 2016: 187–188.

Fig. 23: Gathering Eggs, *Tacuinum sanitatis*, Bibliothèque nationale de France, nouvelle acquisition latine 1673, f. 60 recto. Source: gallica.bnf.fr.

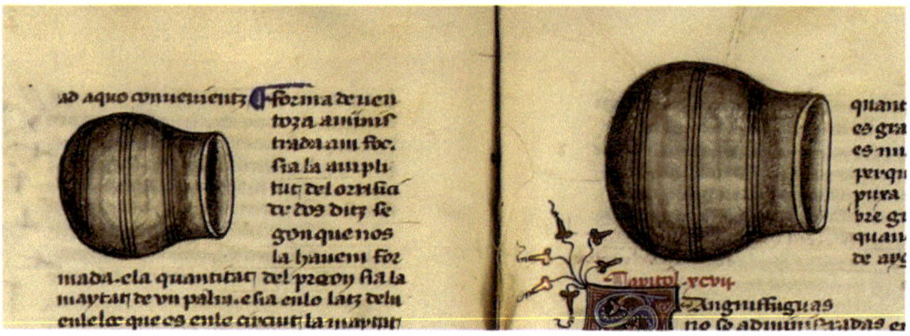

Fig. 24: Tools for Cupping, Albucasis, *Chirurgia*, Montpellier, Bibliothèque Universitaire Historique de Médecine, H 95, ff. 55 verso–56 recto. Université de Montpellier, Crédit photographique SCDI – Montpellier, Service photographique.

egg preparation? The *Tacuinum sanitatis* appears to recommend moderately cooked eggs over hard-boiled eggs. "Les œufs à la coque nourrissent rapidement et abondamment" ("Soft-boiled eggs nourish quickly and abundantly"); hard-boiled eggs cause constipation.[492] The *Elucidari* devotes a paragraph to the preparation of eggs, identifying four different methods: "ad horas so fritz, alcunas vetz bulhitz, autras vetz assatz, ad horas en suc cuechz" ("fried, boiled, roasted, and cooked in a sauce"); eggs that are soft-boiled appear to be the most beneficial.[493]

Ventosar (l. 352). Cupping therapy is an age-old procedure which involves gently heating glass cups (*Albucasis* suggests metal or horn objects can also be used)[494] and then placing them on bare skin.[495] The belief is that the suction created when the cups are applied will relieve pain or swelling, "Las ventosas atyro'l sanc" ("cupping draws the blood").[496] References to cupping in medieval Occitania are several, including *Albucasis*' medical treatise, the *Elucidari*, and the statutes of the barbers of Toulouse.[497]

Vin (ll. 316, 446). Wine was an essential beverage in medieval Occitania, much preferred to water, beer, or cider.[498] Ibn Butlan describes various kinds of wine, from

492 Ibn Butlan, *Taqwim al-Sihha* (ed. Elkhadem 1990: 175).
493 *Elucidari* (ed. Hershon, Ricketts et al. 2018: 598).
494 "Ventozas so de corns, o de eram, o de veyre", Albucasis, *La Chirurgie d'Albucasis* (ed. Grimaud, rev. Lafont 1995: 204); *La Chirurgia* (ed. Elsheikh 1992: 206).
495 The Occitan *Albucasis* devotes several pages to the topic, starting with "De ventozacio et de las qualitatz de uzar de aquelas", Albucasis, *La Chirurgie d'Albucasis* (ed. Grimaud, rev. Lafont 1985: 204–209); *La Chirurgia* (ed. Elsheikh 1992: 206–212). The Montpellier witness of this Occitan text includes two illustrations of cups, on ff. 55 verso and 56 recto; see Figure 24.
496 This line from the *Elucidari* (ed. Hershon, Ricketts et al. 2018: 124) is quoted by Raynouard 1836–1845: 5.500b, s. v. ventosa, ventoza.
497 These references are all mentioned in Raynouard 1836–1845: 5.500b, s. v. ventosa, ventoza and ventozar.
498 Pfeffer 2016: 113–125 and 2005.

white to red, with differing properties, though all are warm and dry.[499] Arnaldus de Villanova sang the praises of wine as a medicinal remedy for all sorts of ills,

> Le vin convient à tous les ages, suivant l'élaboration, la saison, la région. C'est un remède qui convient aussi bien aux personnes âgées qu'aux personnes jeunes. Aux premières, il apporte l'aide pour lutter contre la sécheresse, tandis qu'il est un véritable aliment pour les secondes. Aux enfants, il apporte la chaleur tout en assèchant la moiteur. Aux femmes, il assèche l'humidité lors des contractions utérines. Les savants en font une thérapeutique majeur.[500]
>
> (Wine is appropriate for all ages, depending on its fabrication, the season, the region. It is a remedy that is appropriate for the elderly and for the young. For the former, it helps fight dryness, whereas it is a real food for the latter. For children, it brings warmth while drying moistness. For women, it dries humidity during uterine contractions. Learnèd men use it as a major therapeutic.)

What the author of *Occitan Health Advice for the Layperson* considered a good wine is impossible to determine. Clearly, though, we know what the author did not like as wine, one that is spoiled and "green", "verduros".

Vin agre (l. 369). Ibn Butlan describes vinegar as cold to the first degree and dry to the third, best consumed during the summer months;[501] the *Tacuinum sanitatis* also recognizes wine that has begun to turn bitter, distinguished from vinegar, as this second liquid is warm and humid, intended for those with a cold and dry temperament.[502] The *Elucidari* considers vinegar cold, useful for chilling parts of the body, "a refrigerar las partidas extremas es may util vin agre ... quar vin agre es may agut et penetratiu" ("vinegar is very useful to chill the extremities, ... for vinegar is very sharp and penetrating").[503] Vinegar or verjuice, a very acidic liquid made by pressing unripe grapes, was used often in medieval cooking.[504] Food historian Barbara Santich notes that vinegar and verjuice were interchangeable ingredients,[505] a statement confirmed in the Montpellier medical recipe collection, where the directions in a recipe for fever say to cook purslane in vinegar or in verjuice.[506]

---

499 Ibn Butlan, *Taqwim al-Sihha* (ed. Elkhadem 1990: 204).
500 Arnaldus de Villanova, *Le Livre des vins* (transl. Gifreu 2011: 46).
501 Ibn Butlan, *Taqwim al-Sihha* (ed. Elkhadem 1990: 168).
502 Ibn Butlan, *Taqwim al-Sihha* (ed. Elkhadem 1990: 204).
503 *Elucidari* (ed. Hershon, Ricketts et al. 2018: 487).
504 The *Modus* calls for verjuice, "agras", in four of its recipes, numbers 2, 37, 39, and 41. Recipe number 7, titled "Agras", is actually a recipe for a sour sauce; "agras" is not incorporated into this recipe. See also Adamson 2004: 28.
505 Santich 2018: 29.
506 Brunel 1957: 301–302, recipe number 6d.

Fig. 25: Drinking Wine, *Tacuinum sanitatis*, Bibliothèque nationale de France, nouvelle acquisition latine 1673, f. 77 recto. Source: gallica.bnf.fr.

Local Color: *Occitan Health Advice* Compared to the *Epistola*

If we compare the foodstuffs mentioned in the *Epistola ad Alexandrum de dieta servanda* with foodstuffs mentioned in *Occitan Health Advice for the Layperson*, we observe that the Occitan text contains almost three times as many items specified and that the two texts name the same items in only thirteen cases. This information offers a clear example of how the Occitan author augmented his text, providing much more specific dietary advice than was found in this Latin source.

It is curious, however, that a text composed in Languedoc makes almost no mention of olives or olive oil, though these products were certainly in use in thirteenth-century southern France.[507] The only place where olives appear in *Occitan Health Advice* is with reference to brushing teeth; the Occitan text recommends using olive bark (l. 69) where its Latin source suggested a less specific aromatic bark, "corticibus aromaticis". Here the Occitan author certainly inserted a local product into his lines.

In a similar vein, though the text seems to originate near or not far from the Mediterranean coast – Nîmes, Montpellier, and Narbonne are suggested locations (see Linguistic Conclusions) –, there is very little mention of fish in the Occitan text and none, it must be added, in its Latin source. Given the general presence of fish in the medieval diet, this is a very curious lacuna, but one that may perhaps be explained by the Arabic origin of the material. While the Occitan author details fowl (e.g. chicken, capon, partridges, quail), he inserts fish almost as an afterthought (l. 288). Why the Occitan author did not add fish to the list of foods recommended for months when this source of protein was widely consumed in medieval Europe (ready examples include March because of Lent and December because of Advent) is intriguing; I have no response.

Appreciation of mutton was uneven in medieval Europe. Written and archaeobotanical evidence tells us that mutton was consumed from Aix-en-Provence in Provence (Coulet devotes a chapter to this sole animal)[508] to Rodez in the Massif Central[509] to the small village of Cabaret (Aude) in the Pyrenees.[510] The importance of mutton in the medieval Catalan diet was discussed above. The *Epistola* encourages consumption of the flesh of rams in winter; *Occitan Health Advice* does not specify male sheep and instead recommends two-year-old mutton in summertime (l. 317).

The list of spices included in the two texts suggests that Johannes Hispaniensis may not have paid any attention to the flavoring of food. His text mentions only unspecified herbs, salt, and vinegar as elements that add flavor, whereas *Occitan*

---

507 Pfeffer 2016: 90–100 and 2019.
508 Coulet 1988: 411–449.
509 Catalo et al. 1995: 188.
510 Rodet-Belardi and Forest 2009: 125 and 128.

Health Advice mentions the many spices and herbs that were available in medieval markets, albeit often at high prices, including cloves, cumin, ginger, nutmeg, pepper, saffron, and salt. To these can be added lavender, mint, parsley, and sage, which might grow in a backyard garden. Again, these additions show one element of how the anonymous author of *Occitan Health Advice for the Layperson* augmented his source material with contemporary and not exclusively local elements.

Occitan Advice Literature ("ensenhamens") and Genre Definition

Advice literature in medieval Occitan has solicited a good deal of scholarly comment, helped in this by the fact that some Occitan authors described their works as teaching tools, "ensenhamens". German scholar Johannes Bathe was the first to attempt to define and describe this genre. For him, the didactic character of the works and the insistence on sharing knowledge were the most important features.[511] This knowledge could extend to courtly behavior (seen in works by Arnaut Guilhem de Marsan, Garin lo Brun or Amanieu de Sescas) or instructions of a specialized nature, such as Peire Lunel de Montech's advice on how to serve a lord. For Bathe, didactic works such as the *Ensenhamen de taula*, advice on table manners, was not an ensenhamen less because of its content and more because of its limited objectives.[512]

American scholar Don Monson widened the description of the ensenhamen genre, noting that these texts are addressed to a specific disciple presented in the work, that the works present a rather closed world of master and student.[513] Monson adopted Bathe's distinction between practical and abstract or scholastic ensenhamens,[514] to conclude:

> Le premier trait des ensenhamens, c'est leur didactisme, c'est-à-dire la façon directe et explicite dont s'exprime leur contenu moral. Quant à ce contenu, il est "courtois," c'est-à-dire que, dans le contexte occitan, il doit se définir surtout par rapport à la thématique lyrique ... vers une description de plus en plus détaillée, concrète et "réaliste" d'une part, vers une dialectique de plus en plus rigoureuse, abstraite et synthétique d'autre part.[515]
>
> (The first feature of ensenhamens is their didacticism, that is, the direct and explicit manner by which their moral content is expressed. As for this content, it is "courtly", that is, in the Occitan context, it is defined by relation to lyric themes ... towards a description ever more detailed, concrete, and realistic on the one hand, towards a dialectic ever more rigorous, abstract, and synthetic on the other.)

511 Bathe 1904 quoted by Monson 1981: 42.
512 Monson 1981: 44.
513 Monson 1981: 60 and 81.
514 Bathe 1904 quoted by Monson 1981: 81.
515 Monson 1981: 80.

Thus, Monson distinguishes ensenhamens from other Occitan didactic texts, insisting on the courtly elements of the former.[516] He accepts that a medieval author could use the themes, techniques, and even quotations drawn from ensenhamens without composing a work that met his definition of the ensenhamen genre; he uses as an example the *Breviari d'amor* of Matfre Ermengaud.[517]

The major distinction, according to Monson, is this:

> Les ensenhamens se distinguent de tous les autres genres didactiques par le caractère foncièrement courtois de leur didactisme. On y trouve une doctrine courtoise très générale qui tourne habituellement autour de l'amour.[518]
>
> (Ensenhamens distinguish themselves from other didactic genres by the deeply courtly character of their didacticism. We find in them a courtly doctrine that turns habitually around the topic of love.)

According to this definition, *Occitan Health Advice for the Layperson* is not an ensenhamen, even though it matches the formal description of Monson:[519] exordium (ll. 1–43), development of advice (ll. 44–454),[520] brief conclusion (ll. 455–484). The presence of a disciple required by Monson is explicit in our text (see lines 29 and 413),[521] even though this student is more or less fictitious. The anonymous author presents Galen as Alexander the Great's teacher, transmitting advice attributed to Hippocrates.

Occitan Health Advice veers away from Monson's description when it comes to what the American scholar calls "didactisme courtois", courtly didacticism, which he considers integral to the genre of ensenhamen. Our text seems to avoid the courtly world, even as it advises a noble on conduct in what must be that setting. The word "courtly" is used but once in the text, almost surreptitiously (l. 236). *Occitan Health Advice* is in a different world, one closer to the schools than to the courts.

We can conclude that *Occitan Health Advice* is not an ensenhamen as scholars have defined the genre, and, significantly, the word "ensenhamen" is never used in the work. Can it be connected to another medieval Occitan literary genre?

I considered matching *Occitan Health Advice* to the genre of "dit", an Old French literary genre that was tremendously popular in the thirteenth century. The opening lines of our text, "Qui vol auzir un bon tractat / qui iei novelament trobat" (ll. 1–2)

---

516 Monson 1981: 128 and 165.

517 Monson 1981: 149.

518 Monson 1981: 171.

519 Again, according to Monson 1981: 169, "Par leur composition, les ensenhamens présentent un plan caractéristique: ils s'ouvrent par un exorde, ils développent ensuite les conseils, et ils se terminent par une brève conclusion."

520 According to Monson, "Les ensenhamens se distinguent de tous les autres genres courtois par le caractère direct, explicite et soutenu de leur didactisme," Monson 1981: 170.

521 See Monson 1981: 170, "L'instruction se présente comme un discours à la première personne, la voix unique du maître qui enseigne".

recall many of the texts studied by Monique Léonard in her important work on the genre. Her definition of this genre is remarkably broad:

> Le 'dit' est une pièce en vers, non chantée, plutôt brève et construite selon de multiples schémas métriques parmi lesquels certains semblent privilégiés, notamment l'octosyllabe à rimes plates.[522]
>
> (The 'dit' is a work in verse, not sung, on the short side and constructed using a variety of metrical schemes, among which some are preferred, especially octosyllabic rhyming couplets.)

The "dit" would find its audience north of the Loire River during the second half of the thirteenth and the beginning of the fourteenth centuries. A tremendous number of texts were composed, by authors known and unknown, that speak to a wide range of topics. Poets sought to reveal some truth, to present a position statement;[523] the genre was initially intended for oral presentation.

Léonard identified fifteen different types of "dits",[524] including those she called "various teachings", "enseignements divers", into which group we could place *Occitan Health Advice.* However, the dit did not have the same popularity in the Occitan world as it did in French; one would be hard pressed to find examples composed in Occitan.[525] Furthermore, it is unlikely that the anonymous author of *Occitan Health Advice* was familiar with this northern French genre, far from what appear to be his medical and health interests.

It might be more useful to recognize that *Occitan Health Advice* cannot be associated with a specific literary genre in that language, that it forms part of a large set of didactic works in Occitan. Attempting to see what this corpus consists of, considering only works with a didactic or pedagogic vision, at a minimum, I count these topics:

- arts of poetry, on poetic composition
- alchemical works,[526] such as the *Obratge dels philosophes,*[527] the *Rosari,*[528] and the *Soma* de Bernat Peyre[529]
- medical texts, including the translation of *Chirurgie* of Albucasis, *Occitan Health Advice,* and the *Thesaurus pauperum*

---

522 Léonard 1996: 73.

523 Léonard 1996: 191

524 Léonard enumerates: "dits moraux"; "allégories morales"; "récits moraux" – narrative works where one can find a lesson in the prologue or epilogue, the essential being to tell a moral story – "dits satiriques"; "dits religieux" which are close to "dits moraux"; "récits religieux"; "prières"; "dits de jongleurs", amusing pieces for the crowd; fabliaux; *dits* of "amour courtois" i.e. inspired by the court and/or by courtesy; "dits historiques" which discuss contemporary issues; "éloges"; "enseignements divers"; "poésies personnelles"; and "divers", because they are unclassifiable, Léonard 1996: 217–218.

525 I explore the development of the "dit" in the Occitan context in Pfeffer and Burrell 2022: 20–22.

526 See Thiolier-Méjean 1999.

527 In Thiolier-Méjean 1999: 161–293.

528 See *Lo Rosari* (ed. Calvet 1997).

529 In Thiolier-Méjean 1999: 294–323.

- veterinary texts
- moral instruction
- grammars
- works on etiquette and manners
- encyclopedias,[530] such as the *Elucidari de la proprietatz de totas res naturals*, the *Libre de Sidrac*, and the *Mervilhas de la terra de Ybernia*

The list is long and does not lend itself to a single or simple definition. Even within the categories enumerated here, there is a good deal of variation. *Occitan Health Advice* falls with medical texts; its author wrote a work in verse, hard to compare to the Occitan prose *Chirurgie* based on Albucasis, where the manuscript includes illustrations that mark its potential readers as practitioners, not laypeople. Health is the common theme, but the presentation of information is not at all similar and makes comparisons difficult.

It is possible that the author of *Occitan Health Advice* tells us himself what he thought he was writing. He announces in line 1 that he is presenting a "tractat", a treatise.[531] We should take notice, because the word "tractat" is not found that often in medieval Occitan usage.

Using tools such as the *Concordance de l'occitan médiévale* 2005, we see quickly in the medieval Occitan verse corpus that "tractat" is used more often by Matfre Ermengaud (1245–1322) than by any other author, throughout his masterpiece, the *Breviari d'amor.* In that work, we find "Comensem donc nostre tractat / e nom de sancta Trinitat" ("We begin then our treatise in the name of the Trinity").[532] He announces his text: "Aquest tractatz es de la natura e de les / proprietatatz del albre d'amor" ("This treatise is on the nature and properties of the tree of love")[533] and repeats at the end: "per qu'es mout perilhos auzir / los seguens tractatz o legir" ("Therefore it is very perilous to hear or read the following treatise" [for it may harm the man who reads it]),[534] giving this title to the final section: "Le perilhos tractatz d'amor / de donas" ("The perilous treatise about love of ladies").[535] Nothing could be clearer: for Matfre, "tractat" means treatise.

Another example of this word's use, again using the *Concordance de l'occitan médiéval* 2005, is found in an anonymous work, "The Names of the Mother of God", found in a single manuscript, Bibliothèque nationale de France, nouvelle acquisition française 4232, which may have originated in Touget, a town in the Gers; this section

---

530 Recent scholarship suggests that "books of knowledge" might be a better global term for works such as these. See Williams 2021, in particular.

531 Raynouard 1836–1845: 395 s. v. tractat. Raynouard defines the word as "traité, dissertation, accord".

532 Matfre Ermengaud, *Breviari 2* (ed. Ricketts 1989: ll. 259–260).

533 Matfre Ermengaud, *Breviari 2* (ed. Ricketts 1989: ll. 378T.1 and 378T.2).

534 Matfre Ermengaud, *Breviari 5* (ed. Ricketts 1976: ll. 27641–27642).

535 Matfre Ermengaud, *Breviari 5* (ed. Ricketts 1976: ll. 27790T.1–27790T.2).

of the manuscript may be dated to the first half of the fourteenth century.[536] The anonymous author of this short, fragmentary text, uses "tractat" to describe his effort, at its beginning and at the end: "vos diray un tractat / dels sieus benignes noms que de lieys ay trobat" ("I will recite a treatise of her blessed names that I have found") (ll. 31–32); "En tal maniera ay mon tractat fenit / dels noms de la mayre de Dieu" ("In this fashion have I ended my treatise on the names of the mother of God") (ll. 286–287).[537]

An online resource, the *Trésor Manuscrit de l'Ancien Occitan* 2017– (known as TMAO), offers additional examples of use of the term "tractatus", though the majority represent a verb rather than the noun discussed here. We find "tractat" used in the *Chirurgie* of Albucasis, another medical work.[538] This translation into Occitan was completed for Gaston Febus, count of Foix, who died in 1391. In the *Chirurgie*, we find "tractatus" used repeatedly, with no doubt as to its meaning of treatise:

> Ayssi comensan las paraulas de Albucasim. O, filhs, pus yeu he a vos complit aquest libre, le qual es le derier de la sciencia de medicina, per le compliment de lu so consequit la fi en lu, e per las exposicios de lu, e per las declaracios de lu, es a mi vist que yeu complesca aquel a vos am aquest tractat, le qual es partida de la operacio am ma, so es cyrurgia.[539]
>
> (Here begin the words of Albucasis. Oh, son, since I have completed this book, which is the last in the science of medicine, for the complement of what follows to the end and for the exposition and the declarations of it, it is clear to me that I complete this treatise for you, which discusses the part on manual operations, that is, surgery.)

The author returns to the word later in his work: "E aprop ras le cap tot del malaute e pausa sus le coguot de lu emplaustre de cetre descrit en le tractat dels emplastres ("And afterwards, shave the head of the sick person and place the poultice on the cuckold [sic], as described in the treatise on poultices").[540]

Wartburg 1922–2002 includes TRACTATUS among his etymons and points to Occitan for an early vernacular use,[541] though without including nearly as many examples as here. He does add an early fourteenth-century text, Raimon Feraut's *Vida de Sant*

536 "The Names of the Mother of God", also known as "Le traité des noms de la mère de Dieu", is in a section of the codex where we find an owner name, Arnadi Glibi (Paul Meyer suggests these letters represent Arnaut Guilaberti, see "Traité des noms de la mère de Dieu", ed. Meyer 1880: xc), the date of 1345, and the place name Touget (f. 5 recto); Touget is 50 km to the west of Toulouse.

537 "Traité des noms de la mère de Dieu" (ed. Meyer 1880: civ and cvii). Note that Meyer spotted a missing segment in the middle of this text; his line numbers run from 1–91 and then from 253–289, based on his calculations of how many lines might have been lost, see "Traité des noms de la mère de Dieu" (ed. Meyer 1880: cvi).

538 Montpellier, Bibliothèque Universitaire Historique de Médecine, Université de Montpellier H 95. See Albucasis, *La Chirurgia* (ed. Elsheikh 1992) and Albucasis, *La Chirurgie d'Albucasis* (ed. Grimaud, rev. Lafont 1995).

539 Albucasis, *La Chirurgia* (ed. Elsheikh 1992: 3).

540 Albucasis, *La Chirurgia* (ed. Elsheikh 1992: 14).

541 Wartburg 1922–2002: s. v. TRACTATUS, 9. 143.

Honorat,[542] where we find the term used once to mean "treatise": "si con vos escriuray en lo redier tractat, / el libre c'ausires, cant l'auray romanzat" ("As I wrote you in the last treatise, you will hear in the book when I have translated it") (ll. 4124–4125).[543] More often, Raimon uses "tractat" as the past participle of the verb "tractar", "to discuss, to treat".

This brief history of the word "tractat", "treatise", suggests that the word moved into the Occitan vernacular from the medical center of Montpellier; I believe its appearance in *Occitan Health Advice* is its first usage outside of Latin. The term was soon adopted by other authors, notably Matfre Ermengaud whose training in law may have exposed him to the word in Latin.[544] Use then spread through the regions of Languedoc and Provence. While not a genre per se, a treatise is clearly what our author sought to write, which suggests a somewhat learned focus for him and for his intended readers.

Medical and Dietary Advice

Scholars of medieval cuisine recognize that the line between dietary advice and medical counsel are not cleanly drawn, neither in the Middle Ages nor in our own day.[545] Culinary historian Bruno Laurioux has argued that there was no medical influence at all on medieval cuisine.[546] He sees the development of dietary literature as coming from a medical and scientific interest in food[547] and identifies two different currents that meet in the dietary texts: literature devoted to health, which leads to the regimen-of-health corpus, and collections of medical recipes, in which doctors or pharmacists might discuss how to treat patients using foodstuffs. Neither of these groups of texts includes cookbooks and recipe collections for the kitchen, rather than the pharmacy. Other food historians do not agree fully with the French historian.[548]

542 I use the Occitan spelling of this author's name here and in the list of Sources. However, we must note that Ingegärd Suwe used a somewhat curious French spelling, Raimond Feraud, in the edition cited.

543 Quoted in the *Concordance de l'occitan médiéval* 2005. See Raimond Feraud, *La Vida de sant Honorat* (ed. Suwe 1943).

544 See Hershon 2023: 99.

545 See Bénézet 1999: 333.

546 Laurioux 2011: 56.

547 Laurioux 2011: 57. Coming from a different center of interest, Marilyn Nicoud believes that "régimes de santé" respond "to the needs and expectations of the powerful" (Nicoud 2014: 178). She too does not tie the regimen literature to culinary interest. Despite the title of the work, Hofmeister-Winter et al. 2014, *Der Koch is der bessere Arzt: von Diätetik und Kulinarik im Mittelalter und in der Frühen Neuzeit*, the contributors to the volume in German do not touch on this subject either.

548 See Pfeffer 2016: 182, for additional approaches to this question.

Though food scholars have tended to place *Occitan Health Advice* with the food corpus – its earliest name has certainly played a factor in this decision – its contents suggest it would more properly fit in the regimen-of-health genre, of which it may be the only Occitan example. The inclusion of advice found in diet calendars, texts that are clearly medical in intent, and using the months to present this information, seconds this statement.

Given the Galenic assumption that humoral unbalance led to illness and that balancing humors in the human body was a means to health, it seems to me that medical professionals must have thought that food and diet were a mechanism to maintain health, even if these doctors and pharmacists did not work in the kitchen. Since the proper balance of foodstuffs was thought to help accomplish the goal of good health, diet became part of medical treatment.

At the medieval table, humoral theory underlay much of medieval culinary thinking, as individuals sought to consume foods that would balance their natures, or the time of year, or their environment, combining foods that are "dry" with foods that are "wet" to achieve dietary equilibrium and good health.

Though *Occitan Health Advice* does not emphasize humoral theory in its culinary advice, a balance of humors is implied in much of its dietary counsel. For example, our text discourages consumption of water as a beverage, "it extinguishes natural heat", it is suicidal (ll. 203–210). Even when used to take an electuary, the advice is to drink very little, a "glob" (l. 98). We find parallel thinking in the *Livre de Sydrac*, which encourages drinking wine, not water:

> Lo vins es una preciosa causa e digna, et es a salut dels cors e de l'arma, car per lo vin pot om salvar son cors de moltas malaudias e s'arma d'issament; car vins es bons per las savias gentz que lo bevon atrempadament et a razon, e non fan per lhui nulh damnatge a lor ni ad autra gent: per que atals gentz aprofiecha mais trop beure vins que aiga.[549]
>
> (Wine is something noble and precious which is healthy for the body and for the soul; thanks to wine one can heal the body of numerous illnesses and heal the soul in the same way. For wine is good for wise people who drink it with sobriety and intelligence, and they do not harm themsevles nor others; for these people, it is better to drink wine than water.)

We see, therefore, that *Occitan Health Advice* demonstrates that dietary advice and medical counsel could be one and the same.

While *Occitan Health Advice* advises on when to perform blood-letting, its advice is not nearly as detailed as contemporary texts devoted specifically to that topic or to surgical treatments in general, such as the *Chirurgie d'Albucasis*. This lack of specifics suggests the author's audience is the general reader. One might compare the medieval texts to twenty-first century sources: the difference between looking up "phlebotomy" on Wikipedia.org and reading *WHO Guidelines on Drawing Blood: Best Practices in*

549 *Livre de Sydrac*: f. 101 recto; Bec 1987: 2.162.

Phlebotomy, available on the United States National Institutes of Health (NIH) and World Health Organization (WHO) web sites.[550] The first offers information for a non-specialist audience, the second addresses professionals in the field.

Medical advice can be conveyed both formally, as in the *Occitan Health Advice* verses, where octosyllabic couplets suggest careful formal composition, and informally. An example of informal transmission appears in a Latin letter attributed to Petrus Fagarola of Valencia, written for his sons studying in Toulouse in 1315.[551] While the letter mentions two sons, the advice is almost entirely directed to a single individual; the letter may have been meant for "an indefinite general reader or member of a student body", i.e., a broader audience.[552] Petrus's advice covering a range of health issues has some parallels to counsel offered in *Occitan Health Advice*, for example, "Cave dormire ressupinus nisi raro, ... sed sompnus tuus sit in latere aut super ventre et primo super latere dextro deinde super sinistro" ("Avoid sleeping on your back except rarely, ... but sleep on your side or stomach, and first on the right side, then on the left"),[553] which connects directly to lines 227–230:

> E dormiras el dextre las
> un pauc, e pueis ti giraras
> el senestre autre petit,
> tro c'aias ton dormir complit.
>
> (You will sleep on your right side
> a bit, and then you will turn
> on the left for a bit more,
> until you have finished your nap.)

It is, of course, entirely possible that this specific tip was generally accepted and should not surprise. The letter serves, though, to show the common theme of health advice in different genres and outside the formal discipline of medicine.

Culinary Context

For culinary historians, cookbooks of the period being studied are an important primary source. The Middle Ages produced a significant number of cookbooks in most of the languages of the period.[554] Individual recipe collections in Old and Middle

---

550 World Health Organization 2010.

551 Petrus Fagarola, "Advice from a Physician to His Sons" (ed. Thorndike 1931).

552 Petrus Fagarola, "Advice from a Physician to His Sons" (ed. Thorndike 1931: 111).

553 Petrus Fagarola, "Advice from a Physician to His Sons" (ed. Thorndike 1931: 12); Petrus Fagarola, "Translation of a Letter from a Physician of Valencia to His Two Sons Studying at Toulouse 1315 A.D" (transl. Thorndike 1931: 17–18).

554 See Lambert 1992: 317–362 for a listing of 133 medieval cookbook manuscripts or fragments in the collections of libraries from Aberystwyth to Wolfenbüttel, known at the time of publication of

French, Middle English, and medieval German have been analyzed in depth, as these texts tell us, inter alia, what raw materials were used, what dishes were appreciated, and how tastes changed over time.[555]

In comparison to these three language areas, Occitania is a poor cousin. We have only one extant collection of recipes from the region, a text with the title of *Modus viaticorum preparandorum et salsarum*, found in Paris manuscript, Bibliothèque nationale de France, fonds latin 8435, ff. 24 verso–28 verso, and known as the *Modus*; this work is not yet generally available in a critical edition, though Bruno Laurioux has plans to publish Carole Lambert's unpublished research on this text.[556] The language of this text has been the subject of some scholarly debate; I believe it should be considered a Latin text with heavy Occitan influence.[557] Lambert argues that the *Modus* manuscript was copied in Languedoc on paper associated with the cities of Toulouse, Rodez, and Montpellier;[558] Marie-Claire Gérard-Zai dates the *Modus* to 1380–1390.[559]

As a recipe collection, the *Modus* is typical of works of the period: very succinct and not readily used by individuals without a good deal of culinary expertise. While we can call the *Modus* a cookbook, its fifty-one recipes come closer to what the French call an "aide-mémoire": a list of ingredients and perhaps one or two directions as to preparation. Here is a sample recipe, pulled from the collection:

> Si vis facere limonieyra blanca decoque pullos vel gallum cum carne salsa et distempera de amicdala cum brod[io]. Deinde distempera de ris tembratum et moltum et frezis pullo[s] semipartitos cum pinguedine funduta. Recipe lac fac bullire et custodi ne cremetur. Et pone satis de sucre vel unum parum d'agra[s] vel satis de aqua de limos. Et pullos superpone et dona eos cum satis de sucre.[560]
>
> (If you wish to make a white limonieyra,[561] cook chickens or a cock with salted meat [bacon]. Mix almonds with bouillon and a lot of soaked rice. Divide the chickens in two and sauté them in melted grease. Boil some milk, taking care that it not burn. Add sugar or a bit of verjuice[562] or sufficient lemon juice. Put the chicken on top [on the sauce] and serve with sufficient sugar.)

Visible in this recipe are elements of medieval Occitan cookery: the use of almonds, rice, sugar, and lemon juice, very specifically. No quantities of ingredients are pro-

---

that book. As a result of the CoReMA Project, fifty-eight different medieval German-language cookbooks have been made available on line.

555 See Adamson 2002, the essays in Bruegel and Laurioux 2002, and Laurioux 2005, inter alia.

556 Lambert 1989b. See also Lambert 1989a, 1992, 1997–1998, and 2002.

557 Pfeffer 2019. See also Gérard-Zai 2009 on same topic.

558 Lambert 1989b: 39.

559 Gérard-Zai 2009: 531.

560 Lambert 1989b: 151–152; recipe 41, *Modus*: f. 27 verso.

561 The limonieyra is a dish very typical of Occitania, with its use of lemon juice as the bitter element. One can compare this to the suggestion of meats with vinegar (ll. 283–284) found in *Occitan Health Advice*.

562 Verjuice is a non-alcoholic grape by-product, used alongside vinegar.

vided; cooking time is left to the cook's expertise, as is most often the case with medieval recipes in any language.

While *Occitan Health Advice* invokes a number of foods, spices, and dishes, few if any of these are distinctively Occitan in their nature. The one foodstuff that may represent the Mediterranean origin of the Occitan text is the "panada" (ll. 335 and 395), which I have translated as empanada.[563] Late medieval Catalan cookbooks offer a number of recipes for empanadas. For example, the *Llibre de totes maneres de potages* has a recipe for "panades de hocels ho de pardius", bird or partridge empanadas.[564] A similar recipe is found in a parallel Catalan recipe collection, the *Libre de Sent Sovi*, which offers a recipe for "panades d'auçells", empanadas filled with fowl's meat.[565] The *Sent Sovi* follows this recipe with another for "panade de lampresa", lamprey.[566] Any of these may be what the anonymous author of the *Occitan Health Advice for the Layperson* had in mind for his audience.

The Matter of Taste

Let us turn to the invocations of taste, "sabor" in Occitan, and its derivatives, "saboros, saboroses, saborozament" in *Occitan Health Advice*. The anonymous author of this work advises his reader to consume food and drink that are tasty, as these will contribute to good health. Words built on the root of "sabor" are used three times in the space of twenty lines during a larger discussion of behavior at the table. The repetition offers a strong suggestion of the importance of taste.

> E manja *saborozament*
> d'aquo c'auras mais de talent
> a bel pan de froment triat,
> car sapias en veritat
> que tota via t'er plus sa
> so que melhor *sabor* t'aura
> si vols al cor son plaser far
> que si negun de ton menjar
> mais aisso garda solament,
> que, si trastuit aquel pulment –
> eran durs e l'us era mols –
> non fos mingha lo cuex tant fols,
> que·l mol ti serves a derrier,

563 Note that Old French recipes for similar dishes call them *pastés*, "pâtés"; see, e.g., "pastés de connis", recipe number 83 in Redon et al. 1998: 242 and "pastés de veau", recipe number 86 in Redon et al. 1998: 243.
564 *Libre de Sent Soví* (ed. Grewe 2004: 174), recipe number 138.
565 *Libre de Sent Soví* (ed. Grewe 2004: 175), recipe number 139.
566 *Libre de Sent Soví* (ed. Grewe 2004: 87), recipe number 48.

c'ans lo ti dones a premier.
Encontra, si la majers partz
(c'aissi o enseinha la artz)
sera mols e l'autre tenens,
aquo dur pren primieiramens
ab vin que sia clars e ros,
ben flayran o ben *saboros*.
(ll. 165–184)

(Eat *delectably* / of what you want most, / with fine bread of sifted wheat, / for you know, truly, / that your whole life will be healthier / because you will have better *flavors* / if you want to please your heart. / Of any of your dishes / avoid only [this] / that if you are offered foods – / some hard and another which is soft – / if your cook is so crazy / that he serves you the soft dish last, / he should give it to you first. / On the other hand, if the greater / (for this is what the science teaches [us]) / is soft and the remainder firm / then take the firm part first / with wine that is clear and red, / flavorful and *good tasting*.)

The advice boils down to "If it tastes good, it's good for you". In addition to the repeated mention of "sabor", we have directions to the cook as to how to present elements of a meal, following humoral rules understood to promote good digestion: that soft or liquid foods should be served before those that are firm or hard.

The list of items considered tasty and appropriate for consumption during the autumn months (notably the foods spiced with ginger or saffron, l. 320) may remind the reader of the fictional recipient of *Occitan Health Advice*, Alexander the Great; audiences may consider ginger and saffron as expensive spices, more available for a king than for the common man. In truth, saffron was grown in parts of Spain and of Occitania during the Middle Ages – it was not as expensive as one might think for residents of these areas.

Adonchas deu hom plus manjar
qu'en estieu; e deu hom usar
caudas causas e humorosas –
e dousetas e saborosas –
aissi com son razim madur
e figuas dousas ab vin pur,
e grasses moutos de dos ans
e pollas e aucels volans
a bon jussel, en qu'om molra
de gigimbre o de safra.
(ll. 311–320)

(Then must one eat more / than in summer; one must consume / warm and juicy things – / sweet and tasty – / such as ripe raisins, / sweet figs with pure wine, / two-year-old fattened mutton, / chickens and wild fowl / with a good sauce, into which one will grind / ginger or saffron.)[567]

567 Compare the *Modus*'s recipe 10, for a somewhat generic sauce: "Ad faciendum salsam gruis vel piscis aut alterius carnis recipe gyngibre pebre safran et de pane gata quod sit tostus neque combustus et de jecure et distempera cum aceta vel cum aqua calida et pone unum parum de musto bullito et

The other foodstuffs mentioned in these lines were readily available in Languedoc, the probable point of origin of this text. This Occitan list of items to consume is longer and more detailed than that of the Latin *Secretum secretorum*, which simply advises eating chicken and lamb, sweet grapes, and aged and subtle wine ("pulli et agni, uve quoque dulces et vinum quoque vetus et subtile"),[568] or the medical calendars which are even more terse.

The opposite of things that taste good, things that taste bad, is also considered in *Occitan Health Advice*, though much less intensively. The physician advises his patient that an early death comes to those who spend their time in sadness and fear, consuming stale bread or wine that is not good. The Harley manuscript focused on the stale bread, "trop dur pan" and offered a reading that is hard to comprehend (see the note to line 445 of the text):

Aquesta mortz ven majorment
ab ira e ab marriment,
ab penssament e ab tristor,
ab temensa, e ab paor,
ab trop dur pan e saboros,
et ab vin cailh e verduros
(Harley L1, its ll. 405–410)

(This [premature] death comes mostly
from anger and bitterness,
from worry and sadness,
from anxiety and fear,
from bread that is too hard and tasty [sic],
from wine that has turned or is green).

The Barberini manuscript, V1, is more explicit, calling stale bread "mal saboros" "bad tasting" (and, I assume, hard to eat):

Aquesta mort ve exament
Per ira o per pensament
Ab marriment o ab tristor
Ab dolor de cor e ab pahor
Ab pan dur mal saboros
E ab vin terbol e verdegos
(V1, its ll. 378–383)

---

cum sucre pro dulcorando" ("To make a sauce for cranes or fish or another meat, take ginger, pepper, saffon, and white bread – toasted, not burned – and the liver. Dilute these with verjuice or with warm water. Add a bit of boiled must and sugar to sweeten it."), in Lambert 1989b: 138; *Modus*: f. 25 recto.

568 Suchier 1883: 478, ll. 123–124. Brinckmann's text offers a slightly different set of items: "pulli et agni, aquae quoque dulces, vinum vetus et subtile et cetera huiusmodi" ("chickens and lamb, somewhat sweet waters, aged, subtle wine and similar items") (Brinckmann 1914: 44, ll. 154–156).

(This [premature] death comes exactly
from anger and from worries,
from bitterness and sadness,
from pains of the heart and from fear,
from hard bread that tastes bad
and from green and cloudy wine).

In other words, what tastes bad is bad for you.

Though *Occitan Health Advice* does not consider taste as a topic for discussion in itself, the examples in its lines allow us a glimpse of what was thought to taste good in fourteenth-century Languedoc. More importantly, the work offers the heart-warming advice, that what tastes good is good for you, along with the parallel encouragement that what tastes bad will lead to an early death.

Conclusions

A fascinating text, this Occitan work combining elements from a number of Latin sources including the *Secretum secretorum* and regimen literature has solicited insufficient interest since scholars first called it to our attention in the late nineteenth century.[569] This careful examination of the work demonstrates how the anonymous Occitan author pulled material from a variety of sources to create a new and original work of medical and culinary counsel. I know of no similar work that combines medical advice – with a focus on foodstuffs as the means to health –, the timing of bloodletting, and the incorporation of medical calendars in a medieval vernacular language. There are no recipes in this text, neither medical nor culinary, but its utility for an individual not trained in medicine is clear; its importance for modern scholars is equally evident. The linguistic analysis suggests the author was from the Languedoc region; he may have access to materials near Avignon (consider the multiple works found in the medical manuscript now in Montpellier that are echoed in the Occitan text); he has left us a work of interest to historians of medieval food, medieval medicine, and medieval culture in general.

German philologist Hermann Suchier published the Harley manuscript in 1883 and a critical edition using all four extant manuscripts of the verse text in 1894;[570] Yole Ruggieri published an edition of the longer of the two Vatican witnesses, Biblioteca Apostolica Vaticana, *Barberiniani latini*, 311, with a brief linguistic introduction in 1930. Bruno Laurioux makes passing reference to it in his article on spices;[571] Carole Lambert cited the *Diététique provençale*, i.e., this *Occitan Health Advice*, in her

569 See, inter alia, Langlois 1889: 303 and 307.
570 *Provenzalische Diätetik* (ed. Suchier 1894).
571 Laurioux 1983: 31.

study of the only extant medieval Occitan cookbook, the *Modus*.[572] Among recent scholars, historian Louis Stouff seems to have used this Occitan text the most, notably in his study of medieval Provençal eating habits. The massive Alexander the Great project devoted to the medieval myth of Alexander directed by Catherine Gaullier-Bougassas is remarkable for its skill in ignoring these Occitan witnesses, mentioned several times in passing but omitted from the general bibliography,[573] even though the king is definitely evoked in *Occitan Health Advice for the Layperson* at lines 413–414, "Alexandri, reis ensenhatz / sobre totz reis aventuratz"; other language versions of the *Secretum secretorum* get a fuller treatment in the project.

This verse text warrants this new edition using Lachmannian methodology to create something of a composite text of 484 lines that will draw the variants together into a coherent whole. The newly edited text gives us a better picture of medieval Occitan interests in food and health, encouraging the integration of this significant text into the scholarly discourse.

572 Lambert 1989.

573 French Agence nationale pour la recherche, project ANR-O9-BLANC-0307-01 (2009–2014). The Occitan texts are barely mentioned in Gaullier-Bougassas 2014, and receive even less notice in the companion volume, Gaullier-Bougassas et al. 2015b.

Part 2: **The Occitan Text and its English Translation**

Presentation of the Edition and Translation

I have transcribed *u* as *v*, *i* as *j* when these letters represent a consonant. There are relatively few abbreviations in the Harley L1 witness. After lines 262 (L1, 246), 344 (L1, 312), 358 (L1, 322), 362 (L1, 326), 370 (L1, 334), 374 (L1, 338), 384 (L1 348), and 392 (L1 356), the Harley scribe marked sections or paragraphs by use of a blank line; I show the blank spaces in the Occitan text and in the translation. Given the Harley scribe's model of marking sections by means of white space, I have used indents throughout the translation to indicate where new "paragraphs" begin.

The text that follows presents the contents of the Harley L1 witness, with lines from Barberini manuscripts V1 and V2 interpolated when these manuscripts complete thoughts or offer additional information. These interpolations can be spotted from the changes in orthography (the two Vatican witnesses have many Catalan traits). I have also inserted two lines of my own composition to complete rhyming couplets. All these additions are marked in the notes. I have marked some words in the translation with an asterisk (*), to signal words discussed or explained in this Introduction, thereby inviting readers to refer back to the section on Vocabulary of Specific Interest.

The translation is generally line by line, with no attempt to replicate the rhymes of the Occitan text.

https://doi.org/10.1515/9783111268125-002

The Occitan Text: *Conselhs occitans de santat*

Qui vol auzir un bon tractat
qui iei novelament trobat,
c'ai traig dels libres ancians
que Ypocras e Galians
escriuseron per rason fina[1]
de la nobla art de medicina,
si veng' entorn[2] mi e auja o
cada uns hom per son gran pro!
El temps del meilhor rey c'anc fos,
d'Alixandri qu'era tant pros –
c'ap donar e ab gailhardia
acabava tot cant volia –
en celh temps reinhet[3] Galians,
tot lo meilhor phisicians,
e plus savis e plus plasens
que anc fos de negunas gens.
De tan gran subtilesa era;
tant fort malaute non trobera,
sol que·l pogues ren far trazir,[4]
non si duptava del guerir.
De sciensia fon tant grans lums[5]
que .clix. volums;
sabem que fes et atrobet
de phisica e·ls enseynet.
e pueis trames per gran honor
de tot cant sabia la flor
en un petit de pargami
al rey, e dis lo enaissi,
"Alexandri, si tu faras
so que en es breu atrobaras

1 V2 offers the variant "atroberon" for "escriuseron".
2 L1 reads "si ieu gen torn", suggesting the scribe was unsure of the meaning. Suchier 1883 prints "*veng' *entorn", omitting "si" at the beginning of the line. Given the number of vowels in the line, it is possible to read the line as an octosyllable beginning with "si"; beyond that, I accept the reconstruction in Suchier 1883.
3 V1 reads "Era viu lo prous".
4 V1 offers "tros quals pogues far res menjar".
5 V1 offers "tant era gran lo seu sabers".

English Translation: *Occitan Health Advice for the Layperson*

He who wants to hear a good treatise
that I have just translated,[6]
I found it in an old book
that Hippocrates and Galen
wrote, with fine thinking,
about the noble art of medicine,
let him come around me and listen
every man for his great benefit!
In the time of the best king who ever was,
the time of Alexander who was so worthy –
through gifts and through good humor
he achieved everything he wanted to –
at that time lived Galen,
the best physician,
the wisest and most pleasing
that ever was in any nation.
He was very clever;
he never met so sick a person who,
unless he could not get him to swallow anything,
he did not doubt that he could cure him.
He brought great light to science
in 159 volumes;
we learn what he did and discovered
about physical health and what he taught.
He then transmitted in great honor
the flower of all he knew
in a small parchment [book][7]
for the king, saying to him,
"Alexander, if you do
what you find in this letter

6 Literally, "found"; "translated" conveys the effort and allows avoiding repetition of the English verb.

7 "Book" must be inferred. Specifying "parchment" implies a volume of some value, as opposed to a book copied on paper.

e gardas lo mieu mandament –
que non lo tengas en nient –
tostemps estaras sans e sals;
e seras quitis de tos mals;
que malautia non venra;
ni ja metjes obs non t'aura
si non per failha de natura
o qualsque mala creatura
no t'aucizia o no·t nafrava
o ab verin no·t poizonava;[8]
c'aquo non pot hom esquivar
mais[9] ab forsa de ben gardar.
l'escrig qu'iei mandi t'enaissi."
Que al levar cascun mati,[10]
quant ti seras ben reveilhatz
ni un petit estendilatz,[11]
ni auras vestit ta camisa[12]
prima, blanca, bela e lisa.
E tu fai ton cap penchenhar,[13]
un pauc escarpir e gratar,
car aquo es grans sanitatz;
que las grossas fumositatz
que son pujades el dormir,
aquo les en fai departir.
Enapres salliras del lieg[14]
alegrament e per delieg.
E puys a cambra exiras[15]
e aço ne oblits pas
tas mans e tos ueills lavaras,
que plus bela lugor n'auras.

8 V1 offers "metzinava".

9 In *Provenzalische Diätetik* (ed. Suchier 1894), the editor prints "neis" though the word is very clear.

10 The prose versions of *Occitan Health Advice* offer similar counsel, starting from this point in the verse text; see notes 153 and 155 in the Presentation, above.

11 L1 reads "esterilatz"; I agree with the emendation in *Provenzalische Diätetik* (ed. Suchier 1894).

12 V1 reads "vestit e calçat". Lines 47–48 replicate advice found in the Montpellier pharmaceutical recipe collection, Brunel 1957: 316–317, recipe number 121.

13 V2 reads "cap gratar". Lines 49–54 replicate advice found in the Montpellier pharmaceutical recipe collection, Brunel 1957: 316, recipe number 120.

14 V1 reads "E puys a cambra exiras".

15 Lines 57–58 are missing in L1 and supplied by V1.

and follow my orders –
do not think of it as nothing –
you will always be healthy and hearty;
you will have no physical ills;
no sickness will come to you;
nor will you need doctors
unless by fault of nature
or some evil creature
should kill you or wound you
or poison you with venom,
for no one can escape these
except by taking great care.
Here is the text I send you."
On rising each morning,
when you are[16] well awake
and have stretched a bit,
you will put on a shirt,
delicate, white, lovely and soft.
You will comb your hair,
part [it], and scratch your head,
for this is very healthy;
for the big vapors
that develop while sleeping,
this makes them go away.
Then you will jump out of bed
lightly and with delight;
and then you will leave the bedroom.
Do not forget
to wash your hands and your eyes
so you will look more bright.*[17]

16 The Occitan uses a future tense here where English prefers the present.
17 The asterisk here and throughout the translation marks Occitan words discussed in the section on Vocabulary of Specific Interest, above.

La boca lava eissamentz[18]
que plus belas n'auras las dens –
en estieu ab aigua fregeta,
et en ivern sia caudeta.
Aquo fai calor confortar[19]
e rent volontat de manjar.
Enapres arbilha e toca
las dens, e mena per la boca
un pauc de rusca d'oliver,
o de vern, o de presseguier,
de salvia,[20] o de gensana
o de qualqu'especia amarana,[21]
c'aquo fai las dens conservar
e fai la lengua miels parlar,
e purga de fleuma·l cervel.
E·l vezer serva clar[22] e bel,
e·l col e·ls brasses e·ls brazos
rent plus grosses e plus carnos.
en aprop tot alegramens[23]
vestiras tos bels vestimens
[sed'],[24] arlatas, e cisclatons
[lo]s[25] plus richs els plus precios
que ya poyran esser trobast
e sien ben de ton agrat.
Aprop per lo cor confortar[26]
tu manja .i. plein culheirat
de lectuari que sera
aitals cant al temps convenra.
En estyeu o vas lo pascor[27]
cant lo temps torna en dousor,[28]

---

18 Lines 61–66 replicate advice found in the Montpellier pharmaceutical recipe collection, Brunel 1957: 316–317, recipe number 121.
19 V1 reads "cos amundar"; V2 reads "lo cor confortar".
20 V1 reads "fenoll".
21 V1 reads "ho qualque petit de megrana".
22 L1 reads "calr", inversion of two letters.
23 Lines 79–84 are missing in L1 and supplied by V2.
24 Initial syllable is missing, the word "seda", "silk", would fit nicely.
25 Initial syllable is missing; "los" works to complete the line.
26 V2 adds "tu ti faras aportar". Lines 85–88 replicate advice found in the Montpellier pharmaceutical recipe collection, Brunel 1957: 316–317, recipe number 121.
27 Use of synecdoche, the Easter holiday representing the season of spring.
28 L1 reads "dorsor"; I agree with the emendation in *Provenzalische Diätetik* (ed. Suchier 1894).

Wash your mouth also
so you will have lovelier teeth –
in summer with cold water
and in winter with warm.
This encourages body heat
and stimulates the appetite.
After getting dressed and brushing
your teeth, run through your mouth
a bit of olive bark
or of alder, or of peach bark,
or sage, or gentian,
or some other bitter spice,
for this helps preserve the teeth
and helps the tongue speak better
and purges phlegm from the brain.[29]
And your sight will be[30] clear and beautiful,
and your neck and upper arms and forearms[31]
made bigger and more muscular.
And then quite happily
you will put on your lovely clothes
of silk, scarlet cloth, and satin
richer and more valuable
than have ever been found,
and they will be pleasing to you.
Then to strengthen your heart
you will eat a large spoonful
of an electuary* which will be
such as is appropriate to the season.
In summer or near Easter
when the weather turns sweet

29 Bondurand 1895: 195, offers "chasse la paresse du cerveau" for this line, interpretation rather than simple translation.
30 The Occitan "servar" can mean "to preserve", the sense intended here.
31 Bondurand 1895: 195 followed Raynouard's definition of "brazos", "fesses, gras des fesses" (*Lexique* 2.247, s. v. brazon) and apologizes for his translation, "Les conventions assez hypocrites de nos moeurs me forcent d'aller chercher le vieux mot 'nages' pour traduire 'brazos', mot du reste fort innocent". Raynouard was wrong, as Levy 1894–1924 demonstrates: 1.162b, s. v. brazon.

e tu prent de sucre rosat,
dyarrodon[32] reubarbizat;
en ivern tauleta muscada
o de fort bona cominada,
de pebre,[33] o de gingibrat
o de bon diantes, muscat.
Ab lo lectuari d'estieu
met .i. glop de l'aigua del rieu.
Ab aquel d'ivern menbre ti
que bevas .i. pauc de bon vi.
En aprop ti fai aportar
d'espetias[34] per bon flairar,
e sia tals l'odoramens;
con al temps sera covinens:
en estieu, entorn lo pascor,
cant lo temps torna[35] en dousor,
de rozas, o de violetas[36]
o de lis o d'autras floretas;[37]
en yvern vas lo calendor,[38]
cant l'aigua gela per freydor,
porta musquet, o aloes
o basme, o notz de cipres
o autra causa ben flairan,
que sia ad aquo semblan.
C'aissi co·l pans ajudamens[39]
es del cors e sos noirimens,
aissi es condutz e fortors[40]
de l'arma li bona odors.
Bona odors, bels vestimens
dona gaug et esbaudimens;
e·l gaug fai home esforsar

Fig. 26: A Gift of Violets, *Tacuinum sanitatis*, Bibliothèque nationale de France, nouvelle acquisition latine 1673, f. 83 verso. Source: gallica.bnf.fr.

32 V2 substitutes "[vin]aigre".
33 V2 reads "de preis".
34 V1 reads "Queacom".
35 V2 reads "regna".
36 V1 reads "violes".
37 V1 reads "De liris e semblants coses".
38 V1 reads "quand fa frescor". Compare "A calendas, lo sant jorn de nadal" ("At calends, the holy day of Christmas"), *Roman d'Arles* (ed. Haupt 2003: l. 1117), and see Regionally Specific Vocabulary, above.
39 V1 and V2 reverse the rhyme words here.
40 V1 reads "conduyt al cors".

you will take rose sugar,
diarrhodon* with rhubarb*;
in winter, tablets of nutmeg
or of very good cumin,*
of pepper, or of candied ginger,*
or of good carnation extract,* or of nutmeg.
For an electuary in summertime
add a sip* of river water;
for that of winter remember to
drink it with a bit of good wine.
Then have brought to you
spices to smell good
and let these be your perfume;
these will be appropriate to the seasons:
in summer, around spring
when the weather turns sweet,
roses, or violets,
or lilies, or other flowers;
in winter, towards Christmastime,[41]
when water freezes from the cold,
bring musk, or aloe,
or balm, or cypress seeds,[42]
or other things that smell good,
that are similar to these.
Thus, as bread is the helper
of the body and its nourishment,
so foods and strong smells
are good odors for the soul.
Good odors, lovely clothing
offer joy and happiness;
joy makes a man work

Fig. 27: Lilies, *Tacuinum sanitatis*, Bibliothèque nationale de France, nouvelle acquisition latine 1673, f. 84 recto. Source: gallica.bnf.fr.

41 See above, Language of the Scribe, for a discussion of the meanings of "calendal" and "calendor".
42 Bondurand 1895: 196, translates this term word for word as "noix de ciprès". Given that cypress trees have cones and seeds, I have used "seeds" rather than "nuts'. Also found in Occitan texts are the words "notz ycherca" or "notz iserca", which refer to grain of paradise. My sense is that cypress is closer to the aromas suggested here than the melegueta pepper, see Pfeffer 2015.

e·l sen creisser e agusar;[43]
e·l sanc fai correr per las venas
e fai resplandir las codenas.
Apres un petit anaras
deforas, e cavalcaras
per carreiras o per camis
o per boscs[44] o per bels jardis,
e auziras cant dels auzels,
per tal qu'en sias plus irnels.[45]
Enapres tu t'en tornaras
en ton palais e discendras,
e seras[46] ab tos cavaliers,
e auziras tos messatgiers
que novelhas t'aportaran
de las terras don ilh venran.
Pueis parlaras ab tos amics,
ab los plus savis e·ls plus rics
e·ls plus plazens[47] que tu auras
e tals que t'aian bon solatz[48]
et aian lo cors avinent
e bela cara e risent.
E ton consell ab ells tindras[49]
e ton regne ordonaras.
E garda c'hom empaginatz
faisse los, ni mal enseynhatz
ni nulha mala creatura
ni homs ab mal'esgardadura
ni negus homs que dol ti fassa
non auze venir en ta plassa.
C'aissi con al cor es afans
de sostenir los faisses grans,
es grans afans homs mal aibitz
a l'arma e als esperitz.

43 V1 reads "crexer e ben regnar".

44 L1 reads "bocs"; I agree with the emendation in *Provenzalische Diätetik* (ed. Suchier 1894).

45 V1 reads "E veuras los arbres novells". The spelling "irnel-" (7 examples in the *Concordance de l'occitan médiéval* 2005) is less frequent than "isnel-" (20 examples in the *Concordance de l'occitan médiéval* 2005) but is an accepted form. I retain L1's line as a *lectio difficilior.*

46 V1 reads "solaçant".

47 V1 reads "pus feels"; V2 reads "plus fisels". I retain L1's word.

48 V2 ends with this line.

49 Lines 143–144 are missing in L1 and supplied by V1.

to increase and sharpen his intelligence;
it makes blood run through his veins
and makes the skin shine.
After a while you will go
outside and ride
over roads and trails
through woods or lovely gardens,
and you will hear birds singing
such that you will be faster.[50]
And then you will return
to your palace and will dismount
and you will be with your knights,
and you will listen to your messengers
who bring you news
from the lands from whence they come.
Then you will speak with your friends
with the wisest, the most valued,
and the most pleasing[51] that you have
and those who offer you useful solace,
who have pleasing bodies
and lovely and happy smiling faces.
You will hold council with them,
and you will govern your realm.
Insure that no gloomy man,
nor badly educated, harm them,
nor any harmful creature
nor a man with an evil eye,
nor any man who wants to cause you pain
dare come into your presence.[52]
For as it is tiring for the heart
to carry a heavy burden,
the man who has bad qualities is an equally great weight
for the soul and for the spirit.

50 Bondurand offers "de manière à en être plus alerte". "Irnel, isnel" clearly carries the sense of speed. See Wartburg 1922–2002: s. v. *SNEL, 17: 159b.

51 Bondurand 1895: 197, offers "faithful", following the Vatican witnesses.

52 "Plassa" usually means "a place, a city square, a city, a place of residence". From these meanings, it seems reasonable to translate the word here as "presence", as in "come into your residence". Bondurand 1895 offers "ose prendre ta place" which does not seem to fully follow the preceding lines.

E puys les mans te lavaras[53]
e al menjar te poseras.
Pueis, segon qu'er acostumat
c'auras de manjar voluntat,
tu iras far un'endemessa.
E cant sera la taula messa,
don t'aiga als mans tos escudiers,
e aport i lo despensiers
de totz los conduitz c'om poira
trobar ni acesmat aura.
E manja saborozament
d'aquo c'auras mais de talent
a bel pan de froment triat,[54]
car sapias en veritat
que tota via[55] t'er plus sa
so que melhor sabor t'aura
si vols al cor son plaser far.[56]
Que si negun de ton menjar
mais aisso garda solament,
que, si trastuit aquel pulment –
eran durs e l'us era mols –
non fos mingha[57] lo cuex tant fols,[58]
que·l mol ti serves a derrier,
c'ans lo ti dones a premier.
Encontra, si la majers partz
(c'aissi o enseinha la artz)[59]
sera mols e l'autre tenens,
aquo dur pren primieiramens
ab vin que sia clars e ros,[60]
ben flayran o ben saboros.

53 Lines 155–156 are missing in L1 and supplied by V1.
54 V1 reads "Bons menjars e pan triat".
55 V1 reads "ora".
56 Lines 171–172 are missing in L1 and supplied by V1.
57 "Mingha", while unusual, is not an impossible spelling for the word more frequently seen as "mica, miga", see Levy 1894–1924: 5.278a, s. v. mica, miga, mia, and minga.
58 V1 reads "No fosses ja de res tan foll".
59 V1 reads "Del dur qu'axiu diu notre art".
60 L1 reads "rots"; V1 reads "bos".

And then you will wash your hands
and prepare to eat.
Then, according to what is customary,
you will want to eat;
you will put yourself in motion.[61]
When the table has been set,
your squires provide water for your hands,[62]
and the major-domos bring
all the foods that one can
find and have been prepared.
Eat delectably
of what you want most,
with fine bread of sifted wheat,
for you know, truly,
that your whole life will be healthier
because you will have better flavors[63]
if you want to please your heart.
Of any of your dishes
avoid only
that if you are offered foods*[64] –
some hard and another which is soft –
if your cook is so crazy
that he serves you the soft dish last,
he should give it to you first.
On the other hand, if the greater part
(for this is what the science teaches [us])
is soft and the remainder firm
then take the firm part first
with wine that is clear and red,
flavorful and good tasting.

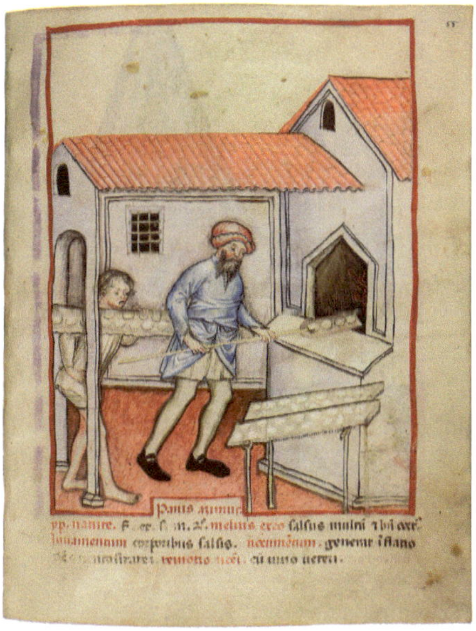

Fig. 28: Baking Bread, *Tacuinum sanitatis*, Bibliothèque nationale de France, nouvelle acquisition latine 1673, f. 55 recto. Source: gallica.bnf.fr.

61 "Endemessa" is an intriguing word, meaning "leap, bound" or "road, way" per Levy 1894–1924: 2.462a, s.v. endemesa and perhaps "limit" per Raynouard 1836–1845: 4.229b, s.v. endemessa, though this meaning is contested by Levy 1894–1924: 2.461a, s.v. endemesa. Emil Levy disputes Suchier's understanding of this line ("wirst du der Unterhaltung ein Ziel setzen", Suchier 1883: 532) and offers "wirst du dir Bewegung machen", see Levy 1894: 403. My translation follows Levy's logic using a contemporary idiom.

62 The verbs in these lines are singular, though the sense is that various servants (pl) will be doing the activities. I translate accordingly.

63 Bondurand 1895: 198, offers "ce qui aura pour toi la meilleure saveur te sera le plus salutaire".

64 Raynouard understands "polmen" as "soupe, potage, ragoût" (*Lexique* 4.592, s.v. polmen) but the sense here is broader than that. Levy 1894–1924: s.v. polmen, 6.433b) understands "Speise, Nahrung" which makes more sense in context. See also Vocabulary of Specific Interest, above.

Mais al manjar cubertamens
fai un pauc de retenimens,
que adoncs t'en sapchas laissar
cant tot poirias mais manjar.
Car ieu sai per fina raison
que trop manjar, contra faisson,[65]
fai home canut e ferran[66]
enant temps e greu e pezant.
E sobrequetot atressi,
non tant lo vespre co·l mati;
d'aisso ti volria castiar,
que·t poiria gran dan donar.
Que pueis non sias trop coichos
de manjar tro·l digestios
sia el ventre celebrada;
et aquo er causa proada:
purgamens, e fams, e saliva
cant er correntz, clara e viva.
Eissamens non vuelhas usar
beure aigua apres manjar,
car en moutas guizas fai mal.
Qu'estien la calor natural.
E cascuns hom qui[67] o faria,
si costumat non o avia,
par si mateix[68] vulla auçir
Son nom e son preu delir
mais si per condutz trop salatz
o per caut eras assedatz,
si que non poguesses sufrir
de beure segon ton albir,
beu d'aigua, al mens que poiras,
la plus frega que trobaras.
Pueissas de la taula[69] moras
e per lo palais anaras

65 V1 reads "menjar ultra saho".

66 V1 reads "canuts en pauchs ans".

67 V1 reads "Tot hom qui"; Suchier prints "so sent cascuns que" in *Provenzalische Diätetik* (ed. Suchier 1894).

68 Ll. 209–210 are missing in L1 and supplied by V1.

69 V1 reads siti.

As for eating to sustain [yourself][70]
restrain yourself
so that you leave some [food]
for when you will be able to eat more.
For I know by careful reasoning
that eating too much, against good practice,
makes a man bald and grey
before his time, and heavy and disagreeable.
And above all this,
not as much in the evening as in the morning
I want to warn you,
for it will cause you great harm.[71]
And then do not be too hurried
to eat while digestion
is completed in the belly;
this is the known process:
laxatives, hunger, and saliva
when they are moving, clear and lively.
Then you do not want to use [or]
drink water after eating,
for in many ways it causes damage.
It extinguishes natural heat.
And every person who does this,
if it is not his habit,
he would want to kill himself
and destroy his name and his worth.
But for dishes that are too salty
or if you are seated too warmly,
so that you cannot wait
to drink according to your wishes,
drink some water, the least you can,
the coldest you can find.
Then you will leave the table
and go through your palace

70 Raynouard 1836–1845: 2.423b, s. v. cubertament, and the *Dictionnaire de l'occitan médiéval*, s. v. cubertamen, both translate "cubertamens" as "in hiding, secretly". "Cubertamen" cannot mean "in hiding" or "secretly" here. Bondurand 1895: 198, translates this line as "Mais au manger, sans affection". If we connect the adverb to the verb "cobrir" which includes in its meanings "to protect, sustain", the line could mean what is presented here: "As for eating to sustain [your appetite, yourself]"; see Wartburg 1922–2002: s. v. COOPĔRĪRE, 2: 1141–1148.

71 The advice is clearly against over-eating. Bondurand 1895: 199, makes this explicit in his translation, "Surtout mange moins le soir que le matin".

sai e lai, un pauc deportan,[72]
ausen versetz, e comandan
a ta gent so que·t plazera[73]
segon lo temps que correra.[74]
E enapres per ton profeig[75]
colguaras ti en ton bel lieg,
garnit e reparat e mol;
e sian prim[76] e blanc li lensol.
E dormiras el dextre las
un pauc, e pueis ti giraras
el senestre autre petit,
tro c'aias ton dormir complit.
Enapres del lieg salliras;[77]
tas mans e tos vuelhs lavaras,
e cant sera tot acabat
so qu'a l'espreisser t'ay mandat,
fa obrir los vueis[78] del palaitz,[79]
e qui sera cortes[80] ni gais
ni adorns[81] ni bos solatiers,[82]
aquel en port meilhors loguiers.
Aquel vuelhas qu'intre primiers,
qu'es de[83] solatz plus plasentiers.
Lains sian li[84] cavalier,
escatz e taulas e taulier
e donzels ab belas colors[85]
que aian garlandas de flors,
juglar ab douces istrumens

72 V1 reads "Say o lay departar t'iras".
73 V1 reads "disen e parlan".
74 V1 reads "Plasers e no trop subtilian".
75 L1 orthography of "profeig" is unique. See The Language of the Author, above.
76 V1 reads "freschs".
77 V1 reads "levaras".
78 V1 reads "portes".
79 Suchier emends to "palais" in *Provenzalische Diätetik* (ed. Suchier 1894).
80 V1 reads "cundes".
81 V1 reads "Adreyt".
82 The *Dictionnaire de l'occitan médiéval*, s. v. solasier, only offers this orthography; Levy 1894–1924: 7.772a, s. v. solasier, includes orthographies of "solascier" and "solacier". Given the related word "solatz", the -t- here does not surprise.
83 L1 reads "que de"; I agree with the emendation in *Provenzalische Diätetik* (ed. Suchier 1894).
84 V1 reads "bons".
85 V1 reads "Donzelles, juglas e scudiers".

here and there, having a good time,
hearing verses and ordering
your people, insofar as it pleases you
according to the moment.
    And then for your benefit
you will take yourself to your fine bed,
decorated and prepared and soft;
the linens will be light-weight[86] and white.
You will sleep on your right side
a bit, and then you will turn
on the left for a bit more,
until you have finished your nap.
    Then you will leave your bed;
you will wash your hands and your eyes,
and when this is all done,
as I have ordered the wake up for you,
you will open the doors of the palace
and he who is courtly and gay,
agreeable and good company,
he will gain the best recompense.[87]
You will have enter first, he
who is the most pleasing company.
Within there are the knights,
chess,* backgammon,* and trictrac,*
and youths of good complexion
who have garlands of flowers,
who play sweet instruments,

86 "Prim" carries meanings that include "first", "delicate", "thin", "subtle", and "meager" (*Dictionnaire de l'occitan médiéval*: s. v. prim). With reference to linens, this could easily mean "light-weight" as I translate here. Bondurand's translation of the term as "fin" (Bondurand 1895: 199) is equally valid.
87 The Occitan text uses a future tense for the description of these courtiers and a present tense for the reward; English usage suggests a different order of tenses.

e juglaressas eissamens.
et aujastz cansonetas belas,
descortz e de[88] baladas novelas
o la gesta o l'estrument,[89]
que a ton con er plus plasent.[90]
E qui sabra miels deportar
ni miels ti sabra dir ni far
per que aias mais d'alegrier,
aquel en port meilhor loguier.
Pueis per lo cor a confortar
fassa hom las trompas[91] sonar,
o, si platz, vai[92] ab ta mainada
un pauc deportar en[93] cassada;
o seras en ost, eissament[94]
fezesses afortidament[95]
ab gran bruig e ab gran esglat[96]
combatre castel o[97] ciutat.

[blank line in Harley manuscript]

Las quatre temporas del an,[98]
non vueil que t'annon oblidan:
co es primaver'[99] e estieus,
automp e yvern ab sas nieus.[100]
Primaveira es plus tempratz
e adoncs es grans sanitatz;[101]
noranta dos jorns son comptats[102]

Fig. 29: Song and Singing, *Tacuinum sanitatis*, Bibliothèque nationale de France, nouvelle acquisition latine 1673, f. 85 verso. Source: gallica.bnf.fr.

88 Suchier omits "de", see *Provenzalische Diätetik* (ed. Suchier 1894).
89 V1 reads "Xanten e sonen struments".
90 V1 reads "Dient vocables molt plasents".
91 V1 reads "Fay les trompes .i. pauch".
92 V1 reads "Per deportar".
93 V1 reads "Puys pots anar a la".
94 V1 reads "Fay en l'ost ton sforcament".
95 V1 reads "E en apres ricosament".
96 V1 reads "brogit e sforçat".
97 V1 reads "De pendre vila o".
98 V1 reads "que son en l'an".
99 Implied is that the calendar year begins at the beginning of spring; it was fairly common in medieval Occitania to begin the year at Easter.
100 V1 reads "ivern frediu".
101 V1 reads "E es d'umor e de saho".
102 Line 269 is missing in L1 and supplied by V1.

and female performers* as well.
And you will hear lovely songs,*
descorts* and new ballads*
with gestures and with instruments,
which will be most pleasant to your heart.
He who knows best how to amuse,
who knows best how to speak or behave
so that you are happiest,
he will get the best recompense.
Then, to comfort your heart,
have trumpets blown,
or, if you wish, go with your household
to amuse yourself a bit in the hunt;
or you will also be prepared for battle
and you should do it with fortitude,[103]
with a lot of noise and great brio,
to attack castles or cities.

[blank line in Harley manuscript]

The four seasons of the year,
I do not want you to forget:
these are spring and summer,
autumn and winter with its snow.
Spring is the most temperate
and therefore is the most healthy;
ninety-two days are counted

Fig. 30: Hunting Boar, *Tacuinum sanitatis*, Bibliothèque nationale de France, nouvelle acquisition latine 1673, f. 91 verso. Source: gallica.bnf.fr.

103 Bondurand 1895: 200, translates this line as "Si tu es à l'armée, fais vigoureusement". He adds, "Ce sont là jeux de prince" (Bondurand 1895: 201).

[las quals son ben utilizatz][104]
de mecinar o de sancnar,[105]
o de belhas domnas baisar,[106]
o de manjar condutz tempratz[107]
que aian bonas qualitatz:[108]
calletas[109] grassas o perditz
e ueos tenres e pols farsitz[110]
e laig de cabra al disnar
e laychuguetas[111] al sopar,
de .ix. kalendes marcii[112]
tro a .xi. kalendes de junii.
En estieu contra[113] la calor
es bona causa de frejor:[114]
ab vin aigre carn de vedel
o de bon cabridet novel,
milgranas o pomas aigretas
e cocombres e cogorletas,[115]
a bon jus vert o ab agras
la carn o·l peis que manjaras.

104 Line 270 is supplied to complete the rhyme.
105 V1 reads "sagnar bo".
106 Line missing in V1.
107 V1 reads "Atresi de bon mejar / e de conduyts te vull mostrar".
108 Line missing in V1.
109 V1 reads "gallines".
110 V1 reads "Polls grossets assats petits".
111 V1 reads "leturgues".
112 Lines 279–280 are missing in L1 and supplied by V1.
113 V1 reads "quant fa".
114 V1 reads "frescor".
115 V1 reads "carabaçes".

[which are well used]
for medical treatment and for blood letting,
or for kissing lovely ladies,
or to eat temperate foods[116]
which have good qualities:
fattened quail or partridges,[117]
tender eggs* or stuffed chicken*,[118]
goat's milk* at dinner time,
and young lettuce*[119] at supper time.
From the ninth of the calends* of March[120]
to the eleventh of the calends* of June.[121]
In summer[122] against the heat,
these are good things[123] for cooling:
veal with vinegar,[124]
or good, young goat* [with vinegar*],
pomegranates, and sour apples,[125]
and cucumbers,[126] and summer squash,*
[seasoning] with parsley[127] and with verjuice*
the meat or fish that you will eat.

Fig. 31: Catching Quail, *Tacuinum sanitatis*, Bibliothèque nationale de France, nouvelle acquisition latine 1673, f. 68 recto. Source: gallica.bnf.fr.

116 While not explicit, the invitation to consume "temperate foods" may be referring to the categorization of foodstuffs in the humoral tradition, which labelled everything on two scales, hot/cold and dry/wet (see Galenic Tradition and Humoral Theory, above).

117 Partridge is ranked as "moderately warm" by the *Tacuinum sanitatis*, Ibn Butlan, *Taqwim al-Sihha* (ed. Elkhadem 1990: 178), which may correspond to "temperate.'

118 Chicken is ranked as "moderately warm" by the *Tacuinum sanitatis*, Ibn Butlan, *Taqwim al-Sihha* (ed. Elkhadem 1990: 180), which may correspond to "temperate".

119 Lettuce is not a temperate food according to the *Tacuinum sanitatis*, Ibn Butlan, *Taqwim al-Sihha* (ed. Elkhadem 1990: 164).

120 The calends are the first day of the month, so the author sets the start of spring on what we would call March 9.

121 We would call this date June 11.

122 The Occitan uses the word for summer, though the context strongly suggests spring is intended.

123 The Occitan uses a singular verb here; in English the plural serves as a better lead into what follows.

124 Vinegar is considered slightly cold and very dry according to the *Tacuinum sanitatis*, Ibn Butlan, *Taqwim al-Sihha* (ed. Elkhadem 1990: 168).

125 Sour or bitter apples are humorily dry and cold in the *Tacuinum sanitatis*, making them a logical food to consume in the summer months, Ibn Butlan, *Taqwim al-Sihha* (ed. Elkhadem 1990: 152).

126 Cucumber is rated moderately cold and moist in the *Tacuinum sanitatis*, Ibn Butlan, *Taqwim al-Sihha* (ed. Elkhadem 1990: 168).

127 The *Dictionnaire de l'occitan médiéval*, s. v. jurvert, is uncertain as to the meaning of "jus vert", glossing "jurvert" as "'verjus (?)' ou 'persil (?)'" and pointing to Levy 1894–1924: 4.284a, s. v. jurvert. It is unlikely that the author of *Occitan Health Advice* would repeat himself in this fashion; in this line, "jus vert" means parsley.

Adonchas no·s deu hom[128] sagnar
ni trob[129] ab donas deportar,
mais ab tot lo meins que poira;
car qui o fai dan i aura.
E deu s'om atressi gardar
en aquel temps de trob manjar.
En l'estiu no haies en ta paria[130]
res[131] qui faça malencolia.
En l'estiu son atresi
huytanta nou dies ço .us afi.
De foch es e de sanitat
axi es lo temps començat
de .xii. kalendes de juny
tros a .ix. de setembre.[132]
Autupne, qui es de fredor,[133]
se es temprada ab humor,
uns terminis es de gran sequor,[134]
naturalmens, e de frejor;
nonanta jorns son complits[135]
e son al .x. jorn exits
a la exida de noembre.
Per ço t'o dich que be.t membre
adonchas deu hom plus manjar
qu'en estieu; e deu hom usar
caudas causas e humorosas[136] –
e dousetas[137] e saborosas –

128 V1 reads "no.t vulles".

129 "Trop" inserted by Suchier to maintain octosyllabic meter, *Provenzalische Diätetik* (ed. Suchier 1894).

130 L1 reads "Segon estieu es la partia / d'automp que fai melencolia". Lines 295–304 missing in L1 and supplied by V1.

131 "Ren, res" can mean "anything" or "anybody" (Levy 1894–1924: 7.224a, s. v. ren; Raynouard 1836–1845: 5.55, s. v. re); the ambiguity is neat and allows the author to advise not only about people but also about things.

132 Ruggieri 1930 and I both transcribe the line as "tros a .ix. de setembre se juny". Ruggieri marked the line with a question mark; I believe the scribe repeated words from the line above. The words "se juny" make the line run long and add nothing to its meaning.

133 L1 reads "naturalmens e de frejor".

134 Line missing in V1; L1 is long by one syllable; Suchier omitted "uns" in *Provenzalische Diätetik* (ed. Suchier 1894).

135 Lines 307–310 are missing in L1 and supplied by V1.

136 V1 reads "amorosos".

137 V1 reads "Ben sanes".

At that time, it is inappropriate to blood let
or to amuse yourself much with women[138]
but rather, [do these] the least you can manage;
for he who does this will be harmed.
And one must take care
not to overeat in that season.
In summer do not have in your company
anyone who causes melancholy.
In summer there are
eighty-nine days, I assure you,
of heat[139] and of health.
This is the season beginning
on the twelfth of the calends* of June[140]
until the ninth [of those] of September.
Autumn,* which is cold,
is tempered by humidity,[141]
a break between great dryness,
naturally, and cold;
ninety days are included
and on the tenth day[142] they are gone
at the end of November.
For this reason I tell you to remember
then must one eat more
than in summer; one must consume
warm and juicy things –
sweet and tasty –

138 Lines 290–291 are omitted by Bondurand 1895: 201.
139 "Fuoc" translates as "fire" but heat (one result of fire) makes better sense in the translation.
140 We would call this date June 12.
141 "Temprada ab humor" can be compared to a line in the *Canso de la Crosada*: "e·l vis de Genestet que·ns tempra la humor" (laisse 163, l. 48). Eugène Martin-Chabot translates this as "le vin de Genestet nous adoucit le tempérament," *Canso de la crosada* (ed. and transl. Martin-Chabot: 2.155); Henri Gougaud proposes "Le vin de Genestet nous fait le coeur content," Gougaud 1989: 279; Claude Mourthé understands the line to mean "Le vin de Genestet tempère nos humeurs," Mourthé 2018: 351.
142 I suspect we are missing something about the calends of December. Logically, the author would start winter on December 10.

aissi com son razim[143] madur
e figuas dousas[144] ab vin pur,
e grasses moutos de dos ans[145]
e pollas e aucels volans
a bon jussel, en qu'om molra[146]
de gigimbre o de safra.
Adonchas deu hom esquivar
cauls e totz legums[147] per manjar,
mais mezinas e purgament
donan adoncs gran leujament;
e adonchas val mais amors[148]
qu'en estieu cant fai grans calors.
Apres ven hyvern ab lo freig
que moutas res[149] ten en destreig.
Adonchas deu hom pron[150] manjar,
e·l cors moure e escalfar,
et esser pres de la cosina;
e manjar tota salvasina;
e far raustir sobre·ls carbons
gallinas, e gails,[151] e capons,
e manjaras raust e panadas,[152]
aste[153] de porc, e carbonadas
que sian trastug salpicat
d'especias e empebrat;
e beu bons vins et bons pigmens[154]
per contrastar als elemens.
E rescon[155] sotz ton cobertor
bela domna ab fresca color –

---

143 V1 reads "rayms".
144 V1 reads "seques e beure purs".
145 V1 reads "Moltons de .ii. dents e bells".
146 V1 reads "Ab la salsa que obs hi ha".
147 L1 reads "lums", which Suchier emended to "liuns", creating a hapax legomenon, *Provenzalische Diätetik* (ed. Suchier 1894). V1 offers "legums", a better emendation.
148 V1 reads "maiors humors".
149 V1 reads "mantes gents".
150 V1 reads "assats".
151 Missing in V1.
152 V1 reads "assats de panades".
153 V1 reads "carn"; Suchier emends to "astes" in *Provenzalische Diätetik* (ed. Suchier 1894).
154 V1 reads "Bevetz los vins clars e pruents".
155 V1 reads "haies".

such as ripe raisins,*
sweet figs* with pure wine,*
two-year-old fattened mutton,*
chickens* and wild fowl*
with a good sauce,* into which one will grind
ginger* or saffron.*
Then one must avoid
eating cabbage* and vegetables,*
but medicines and purging
give great relief at that time;
also at that time love is more worthy[156]
than in summer when it is very hot.
Afterwards comes winter with the cold
which keeps many people[157] in distress.
Then must one eat a lot,
and move and heat the body,
and stay close to the kitchen;
and eat all kinds of game*;
roast* over coals*
chicken* and quail* and capon,*
and you will eat roasts* and empanadas,*
skewers* of pork and grilled meats*
that are completely sprinkled*
with spices and with pepper*;
and drink good wines and good piments*
to contrast with the elements.
Hide under your bedcovers*[158]
a lovely lady with a good complexion –

156 Bondurand 1895: 202, does not translate ll. 325–326.
157 Again, the word "res" can mean "things" or "people" (see note 131, immediately above); note that V1 reads "gents", "people", here. Given that the advice of *Occitan Health Advice* is intended for a person, I translate accordingly.
158 Bondurand 1895 does not translate ll. 341–342.

e non poinhes[159] d'autra mesina
adoncs; que non i a tan fina.

[blank line in Harley manuscript]

Dels .xii. meses eissament
que·t mandi especialment
de quals causas deus estener,
en cascun contra ton voler.[160]
Non ti sancnes[161] en janoier;
ni manges bledas en febrier.
Ni en mars non vuelhas sagnar,[162]
scorradors posar, ni ventosar.[163]
Si donchs gran ops no t'hauia
per cuyta de malaltia
[adoncs non vuelhas consumar][164]
lentilas ni ren dous manjar.[165]
mais sapias que bon seria
ruda, qui beure la podia.[166]

[blank line in Harley manuscript]

En abrilh si deu hom gardar
de tota razitz[167] a manjar,
mais mezin' es bona sangnia
ad aquel que mestier seria.[168]

[blank line in Harley manuscript]

En mai non deu hom cap manjar
de ren, ni es bon de sagnar,

159 V1 reads "curs".
160 V1 reads "En cascu ço que deus voler".
161 V1 reads "No.t vulles sagnar".
162 L1 reads "usar" and jumps to line 356; lines 352–354 supplied by V1, whose rhyme word "sagnar" leads logically to the next medical procedures.
163 Lines 352–354 missing in L1 and supplied by V1.
164 Line supplied to complete rhyme.
165 Missing in V1.
166 V1 reads "qui pendre 'n volia".
167 V1 reads "rael".
168 V1 reads "Es for bo qui ops l'auja".

take no other medicine
at that time, for there is none finer.

[blank line in Harley manuscript]

For each of the twelve months
I order you especially
to avoid these things,
each one despite your desire.
Do not blood let in January;
do not eat chard* in February.
In March you do not want to let blood,*
nor do cauterization,*[169] nor cupping.*
If then you have no great need
for reason of illness
[you will not want to consume]
lentils* or to eat sweet things.
Know that rue* would be good
if you can drink it.

[blank line in Harley manuscript]

In April, one must avoid
eating all root [vegetables],
but it is good medicine to blood let*
him who needs it.

[blank line in Harley manuscript]

In May, one must not eat the head
of anything, nor is it a good [time] for blood letting,*

---

169 The precise meaning of "scorrador", clearly a noun referring to a medical tool, is intriguing. Three modern possibilities come to mind: 1) a cauter, used for cauterization; 2) a tool used for scarification, a medical technique related to blood letting, and 3) a tool to scrape off dead skin, similar to a strigil, used by the Romans to wipe away perspiration. See Vocabulary of Specific Interest above for a detailed discussion of this term.

mais en aquel mes es grans sens
qui usa fenholh ni aussens.[170]
En jung deu hom soven manjar
lachuguas, e deu hom usar
vin agre e tot agriam[171]
a la taula, cant hom a fam.[172]

[blank line in Harley manuscript]

En julh non si fai bon sangnar
ni mezina penre ni dar;
e de las donas fai gran sen
qui no las baisa trop soven.

[blank line in Harley manuscript]

En aost si gart qui savis es
de manjar sobre tota res
causas que·l sanc puesca mesclar
ni colra negra engenrar,
mais causas leus e confortans:
e carns novelhas, e bos pans,[173]
salsa de menta, de polieg.
Es adonchas de gran[174] profieg.
Cel mes non deu hom carn comprar[175]
de bou[176] ni de porc per manjar.

[blank line in Harley manuscript]

Setempbres es tant avinens
que deu far gaug a totas gens.
Car a nuilh home mal non fa,
si en aost pres non o a.[177]

170 V1 reads "encens".
171 V1 reads "agrura".
172 V1 reads "A sa taula qui.l menjar a guha".
173 V1 reads "novelles frequentants".
174 V1 reads "Es ladons menjar de".
175 V1 reads "En aquest mes no deu usar".
176 V1 reads "carn de bou".
177 V1 reads "Si donchs ja abans pres no l'a".

but in this month, it makes good sense
to consume fennel* or absinthe.*
In June one must often eat
lettuce,* and one must consume
vinegar* and all sour fruits*[178]
at the table, when one is hungry.

[blank line in Harley manuscript]

In July it is not good to be bled*
nor to take medicine nor give it;
as for ladies, it makes sense[179]
not to kiss them too often.

[blank line in Harley manuscript]

In August, he who is wise avoids
eating, above all other things,
things that the blood can mix[180]
and that do not create black bile,*
but, rather, light and comforting things:
fresh meat, good breads,
and sauces of mint and of pennyroyal.*
These are of great benefit.
In this month one must not buy
beef or pork for eating.

[blank line in Harley manuscript]

September is so lovely
that it must make everyone joyful.
For it does no man ill,
if he has not been taken [ill] in August.

Fig. 32: Lettuce, Hamilton Siddur, Berlin, Staatsbibliothek zur Berlin, Hamilton 288, f. 23 verso. With permission of the Staatsbibliothek zu Berlin – PK.

178 Bondurand 1895: 203, confuses "agriam" with "agras" here, so he translates the word as "verjus". Levy 1894–1924: 1.35a, s. v. agriam, translates the term as "saure Frucht" following Suchier, who connected "agrium" to "agrum, eigrun", "bitter fruit", Suchier 1883: 639, Glossary, s. v. agriam.

179 Bondurand 1895 does not translate ll. 373–374.

180 Bondurand 1895: 203, translates this line as "ce qui peut brouiller le sang", "things that can mix the blood".

E pot hom miels manjar ses dan[181]
tot condug[182] que en mes del an.[183]
E es bon de mezin' a penre
e de laig de vaca a beure.

[blank line in Harley manuscript]

En oichuvre deu hom usar
rasims douces e most trolhar.[184]
e de tot l'an non es panada[185]
tan sana, si es beu adobada,
e deu hom usar en jussels[186]
pebre, gigimbre e clavels.
En novembre no·s deu banhar
qui se meteis volra gardar,[187]
mais adonchas es[188] profechosa
assatz:[189] sangnia, e ventosa;
et es plus sotilhs a trobar
totz hom qui·s volh' estudiar.
En decembres deu hom tener
de manjar cauls matin e ser,
car podon l'afan engenrar
adoncs, que[190] fai home cridar.
Mais en aquel mes es trop ric,[191]
qui usava[192] l'erba del espic
soven en tot'escudelada,
e gigimbre e fort pebrada.[193]

181 V1 reads "negun dan".
182 V1 reads "Totes coses".
183 L1 reads "anh".
184 L1 reads "trobar"; V1 reads "rayms" and "turullar"; I agree with the emendation to "trolhar", see *Provenzalische Diätetik* (ed. Suchier 1894).
185 Lines 395–396 missing in V1.
186 L1 reads "junssels"; V1 reads "En est mes es bo e bell".
187 V1 reads "matar".
188 V1 reads "Mes en aquest mes es".
189 L1 reads "assa", which might be a *lectio difficilior*; however, V1's term makes better sense in context.
190 V1 reads "per que".
191 V1 reads es "bell e rich".
192 Suchier emends to "usa" in *Provenzalische Diätetik* (ed. Suchier 1894), but it is possible that the words "qui usava" count as 3 syllables and maintain the octosyllable.
193 V1 reads "e bona canyella".

Then one can eat without harm
all foods, better than in any other month of the year.
And it is good to take medicine
and to drink cow's milk.*

[blank line in Harley manuscript]

In October one must consume
sweet raisins,* and one must press must.*[194]
Of all the [months of] the year, an empanada* is not
as healthy [as now], if it is well prepared,
and one must use in sauces*
pepper,* ginger* and cloves.*
In November do not bathe
if you want to protect yourself
but then are sufficiently[195] beneficial
bloodletting and cupping;[196]
and he is more skilled at composing [song],
any man who wants to apply himself.
In December a man must keep
from eating cabbage* morning and night,
for these can then generate suffering*[197]
which makes a man scream.
But in this month he is very wealthy
who uses lavender*[198]
often in his plate,
and ginger,* and very peppery* [foods].

194 "Must", freshly crushed grape juice, containing the skins, seeds, and stems of the fruit, see Vocabulary of Specific Interest, above.

195 I choose to reject L1's reading of "assa", a word Bondurand did not translate (Bondurand 1895: 402). "Assa"* could mean "asafoetida", but that pharmaceutical does not align with the phlebotomy techniques that follow; the Vatican scribe may have better understood the sense of the line than the Harley scribe; V1 reads "asatz", "enough, sufficiently".

196 Cupping is a procedure in alternative medicine that involves using heated cups applied to the skin, creating local suction, see Vocabulary of Specific Interest, above.

197 Bondurand 1895: 204, translates "afan" as "apostèmes", "tumors", which seems a stretch. It is more plausible to understand "afan" as suffering, perhaps resulting from intestinal gas, a frequent by-product of cabbage consumption, see Vocabulary of Specific Interest, above.

198 Bondurand 1895: 204, translates as "l'herbe d'aspic". See Vocabulary of Specific Interest, above, for information on this plant.

Alexandri, reis ensenhatz,[199]
sobre totz reis aventuratz,
menbre·t d'aquo que as auzit
e non o metas en oblit.
Car[200] seynher, que Dieus ti sal.
Garda la calor natural
en ton cor sobre totas res!
Car aitant con sai estaretz[201]
ab humor ses destemprament,
aitant sai vieu hom longament.
Mais pueis hom pren a frevolar,[202]
la vida no·i pot plus estar.[203]
Car en doas guisas si fenis
lo cors de cascun esperitz.
La una es naturalment,
que·s fai per enveilhisement;[204]
cant es vengud' al jorn derrier
natura ses tot destorbier.
L'autra si fai per accident
o per alcun destemprament
qui dampna·l cors[205] per malautia,
enans que hora no seria:
per poiridura o per flairar[206]
o per mals condutz a manjar
per cobeitat[207] o per frachura.
que non troba qui·l pren[208] en cura,[209]
o per colpa, o per naleig,[210]

199 V1 reads "exalçat".
200 V1 reads "bell"; Suchier added "E" at the beginning of the line for meter in *Provenzalische Diätetik* (ed. Suchier 1894); if "Dieus" is counted as three syllables, the line does not need emendation.
201 V1 reads "com aquella y es".
202 V1 reads "baxar".
203 V1 reads "durar".
204 L1 presents the word as "enveilh li sement", suggesting the scribe did not quite understand what he was writing. I adopt the emendation in *Provenzalische Diätetik* (ed. Suchier 1894).
205 V1 reads "Don mor lo cos".
206 V1 reads "Per carns podrides a usar".
207 V1 reads "pobretat".
208 V1 reads "n'aja cura".
209 I do not emend this line and suggest it makes best sense to connect "non" with the verbal expression "prendre cura" rather than with "trobar".
210 L1 reads "nanleig"; V1 reads "O per nafra o colp envellit".

Alexander, learned king,
adventurous above all kings,
remember all that you have heard
and forget none of it.
Dear lord, may God preserve you.
Maintain your natural heat
in your body above all else!
For as long as you will maintain [it]
without perturbation of the humors,
thus, I know, men live long.
But when a man begins to weaken,
life cannot remain in him.

For there are two ways to end
the body of every spirit.
One is naturally,
which happens from aging;
when the last day comes
it is natural without disturbance.
The other happens by accident
or from some perturbation
which ruins the body by illness,
before it would be its hour:
by rottenness, or by smell,
or from eating bad foods
out of greed or out of starvation.[211]
For he who decides not to take care,
either by fault or by negligence,

211 Bondurand 1895: 205, understood "frachura" as "fracture", a broken bone. In this medical text, that reading has its logic; I prefer to connect "frachura" to its meaning of starvation, a situation which would lead people to consume foods that were rotten to assuage their hunger. See Levy 1894–1924: 3.578, s. v. frachura; Raynouard 1836–1845: 3.380a, s. v. frachurar.

que mor malautes en son lieg.[212]
Aquesta mortz ven majorment[213]
ab ira e ab marriment,[214]
ab penssament[215] e ab tristor,
ab temensa,[216] e ab paor,
ab trop dur pan mal saboros,[217]
et ab vin cailh[218] e verduros,[219]
et ab manjar causa ferrada,
e tota causa trob[220] salada,
ab dur jaser[221] et ab veilhar,[222]
et ab trop soven dejunar,
ab beure pols, sutjel[223] o fum,
e de nuitz escriure ab lum,
ab estar torn caitieu fogal,
et ab raïna dins ostal.[224]
Tuit aquist cas ab lor semblantz
abreujan ad home sos ans.[225]
Issament pot hom alonguar
tot so, qui sueng s'en vol donar,

212 L1 reads "sotz to lieg", which Suchier printed as "sotz tolieg", see *Provenzalische Diätetik* (ed. Suchier 1894). I prefer to follow V1 here, as it is slightly more logical.

213 V1 reads "exament".

214 V1 reads "pensament".

215 V1 reads "marriment".

216 V1 reads "ab dolor de cor".

217 V1's reading of "mal saboros" makes better sense than L1's "e saboros". Rejected correction options include connecting the meaning of the word here to a related word, "saboural", meaning salted or rancid meat, or to print "dessaboros" or "dissaboros", building on vocabulary known in medieval Occitan but unattested in this adjectival form. See Matfre Ermengaud, *Breviari 4* (ed. Ricketts with Hershon 2004: 109), l. 18651.

218 V1 reads "terbol".

219 V1 reads "verdegos".

220 V1 reads "o que sia sobres".

221 The *Dictionnaire de l'occitan médiéval* suggests "jasier", "couche, lit", referring to Raynouard 1836–1845: 3.583a, s. v. jasier. Raynouard cites *Girart de Roussillon*, "la nuh, quant gens dormia en son jasier" ("at nighttime when a person sleeps in his bed"). The word "iasier" appears as a variant to l. 9691 of *Girart* (ed. Hackett 1953–1955); it is not included in the *Concordance de l'occitan médiéval* 2005, which used *Girart de Roussillon*, ed. and transl. Combarieu du Grès and Gouiran 1993. See above, Language of the Scribe.

222 L1 reads "vueillar"; V1 reads "dur jaure e trop treballar".

223 "Sutjel" missing in V1.

224 V1 reads "e ab çells de son ostal barellar".

225 Last line in V1.

that sick man dies in his bed.
This death comes mostly
from ire and bitterness,
from worry and from sadness,
from fear and from trepidation,
from bread* that is too hard and bad tasting,
from unfiltered[226] wine* that is green,
from eating offal,[227]
and anything too salty,
from hard beds and from staying up late,
from breaking a fast too often,[228]
from drinks with dust, soot,[229] or smoke [in them],
and from writing at night with [artificial] light,
from being near a weak fire[230]
with disputes in the household.[231]
All these things and matters similar
shorten a person's years.
By the same token, you can lengthen
your life, if you take care,

226 *Dictionnaire de l'occitan médiéval,* s. v. cailh, citing Levy 1894–1924: 1.188b, s. v. calh, suggests that *Occitan Health Advice* means "a fatty wine", "gras [parlant du vin]". Levy's translation of the word in this text is "dickflüssig" suggesting a thick wine; the word "cailh" relates to the process of coagulation, particularly of milk. I believe the author is arguing against an unfiltered wine. Suchier suggested that "vin verduros" was a wine from unripe grapes, Suchier 1883: 645, Glossary s. v. verduros, cited in Levy 1894–1924: 8.663a, s. v. verduros. The variant reading in V1 suggests the line posed problems for a medieval reader.
227 Bondurand 1895: 205, understands this line as "le fer dans les aliments".
228 Bondurand 1895: 205, translates this line as "les repas trop fréquents", the same meaning from a different angle.
229 Bondurand 1895: 205, understood "sueur", "sweat", for "sutjel". The Occitan word is clearly "soot", not "sweat".
230 I understand "a poorly heated environment".
231 Bondurand's translation adds to the Occitan, "un foyer malheureux et une femme acariâtre", Bondurand 1895: 205.

ab gran gaug et ab alegrier,
que[232] totz sos faitz meta primier.
C'ap gaug acaba hom plus gent
totz sos faitz que ab marriment,
car ab gaug si deu hom levar,
et ab gaug deu hom Dieu pregar,
et ab gaug si deu hom vestir
e de bela rauba garnir;
ab gaug deu hom novas contar,[233]
ab gaug ausir et escoutar;
ab gaug et ab alegrament
deu hom portar son vestiment,
et ab gaug deu hom cavalcar
e sos enemics encausar,
et ab gaug anar per son pron
et ab gaug[234] tornar a maison.
Tot gaug e tot alegrament
don' a home confortament,
segurtatz, e pas, e conortz,
trepar, e rire, e deportz
e vezer ribieiras e pratz
e demorar en luecs tempratz[235]
e pron manjar e ben jazer
e gaug de matin e de ser,
tot aisso alongua sa vida
ad aquel qu'aisso non oblida.

232 Suchier prints "qu'a" in *Provenzalische Diätetik* (ed. Suchier 1894).
233 Suchier prints "comtar" in *Provenzalische Diätetik* (ed. Suchier 1894).
234 Suchier inserts "gang", perhaps a misprint, for meter in *Provenzalische Diätetik* (ed. Suchier 1894). I follow his thinking but correct the typographical error.
235 L1 reads "temprat".

with joy and a light heart,
that, in all your activities, you put these first.
With joy, a person accomplishes more courteously
all his deeds than with bitterness,
for one must rise with joy,
and with joy, one must pray to God,
and with joy, one must get dressed
and put on lovely clothing;
with joy, must one tell stories;
with joy, listen and hear,
with joy and with happiness
must one wear his clothing;
and with joy, one must ride
and chase his enemies,
and with joy, take care of his affairs[236]
and with joy, return home.
All joy and all happiness
give a person comfort,
security, peace, and encouragement,
amusement, laughter, and distractions,
and the sight of rivers and fields,
and to remain in temperate places,
and to eat enough and sleep well
in joy, morning and night,
all these things lengthen life
for the person who does not forget this.

236 Bondurand 1895: 206, translates "avec joie doit-on voyager pour son profit". Levy understands "anar per son pro" as "sienem Verdienst, seinen Geschäften nachgehen", Levy 1894–1924: 6.565a, s. v. pro, which the *Dictionnaire de l'occitan médiéval*, s. v. pro_2, translates as "vaquer à ses affaires", to go about one's own business.

Bibliography

Abbreviations

f., ff.	folio, folios
fl.	flourished
l., ll.	line, lines
s. v.	sub verbo

Sources

Albucasis, *La Chirurgia*: Mahmoud Salem Elsheikh (ed.), *Abū'l Qāsim Halaf Ibn 'Abbās az-Zahrāwī detto Albucasis, La Chirurgia, versione occitanica della prima metà del Trecento*. Florence: Zeta, 1992.

Albucasis, *La Chirurgie d'Albucasis*: Jean Grimaud (ed.), Robert Lafont (rev.), *La Chirurgie d'Albucasis (ou Albucasim), texte occitan du XIVe siècle*. Montpellier: Centre d'études occitanes, 1985.

Alfonso X el Sabio, *Libro de los juegos*: Raúl Orellana Calderón (ed.), *Libro de los juegos: Acedrix, dados e tablas, Ordenamento de la tafurerías* (Biblioteca Castro). Madrid: Fundación José Antonio de Castro, 2007.

"Aloes es caut e sec": Matteo Milani (ed.), "'Aloes es caut e sec ...': edizione di un erbario occitano (Firenze, Biblioteca Nazionale Centrale, MS Palatino 856)," *La Parola del testo* 8 (2004), pp. 369–391. Data from TMAO, http://tmao.aieo.org/TrHerbM_show2.

Arnaldus de Villanova, *Antidotarium*: Nicolai Taurelli (ed.), *Arnaldi de Villanova, Opera omnia*. Basel: ex officina Pernea per Conradum Waldkirch, 1585, columns 385–496.

Arnaldus de Villanova, *Le Livre des vins*: Patrick Gifreu (transl.), Arnaut de Villeneuve, *Le Livre des vins*. Perpignan: Éditions de la Merci, 2011.

Arnaldus de Villanova, *Regimen sanitatis ad regem Aragonum*: in Arnaldi de Villanova, *Opera medica omnia* X.1, Luis García-Ballester and Michael R. McVaugh (eds), *Regimen sanitatis ad regem Aragonum*. Barcelona: Seminarium historiae scientiae Barchinone, 1996, pp. 423–470.

Biographies des troubadours: Jean Boutière and Alexander Hermann Schutz (eds), *Biographies des troubadours, textes provençaux des XIIIe et XIVe siècles*, 2nd ed. with collaboration of Irénée-Marcel Cluzel. Paris: A. G. Nizet, 1973.

Bonis Brothers, *Livres de comptes*: Édouard Forestié (ed.), *Les Livres de comptes des frères Bonis, marchands montalbanais du XIVe siècle*, 3 volumes. Paris and Auch: Honoré Champion and Cocharaux frères, 1890–1894.

Brunel 1956: Clovis Brunel (ed.), *Recettes médicales alchimiques et astrologiques du XVe siècle en langue vulgaire des Pyrénées* (Bibliothèque méridionale 1ère série 30). Toulouse: Édouard Privat, 1956.

Brunel 1957: Clovis Brunel (ed.), "Recettes médicales de Montpellier en ancien provençal," *Romania* 78 (1957), pp. 289–327.

Brunel 1966: Clovis Brunel (ed.), "Recettes pharmaceutiques d'Avignon en ancien provençal," *Romania* 87 (1966), pp. 505–542.

Canso de la crosada: Eugène Martin-Chabot (ed. and transl.), *La Chanson de la croisade albigeoise*, 3 volumes (Classiques de l'histoire de France au Moyen Âge). Paris: Belles lettres, 1972.

"Cartulaire d'Albi": Auguste Vidal (ed.), "Cartulaire d'Albi: cartulaire AA1," *Revue des langues romanes* 45 (1902), pp. 447–469.

Chirurgie d'Albucasis: Montpellier, Bibliothèque Universitaire Historique de Médecine, Université de Montpellier, H 95, ff. 1 recto–70 verso.

CoReMA Project: CoReMA Project (Cooking Recipes of the Middle Ages: Corpus, Analysis, Visualisation) directed by Bruno Laurioux and Helmut Klug (ANR-17-CE27-0019-01; FWF Der Wissenschaftsfonds I 2614), https://gams.uni-graz.at/context:corema.

https://doi.org/10.1515/9783111268125-003

Corradini Bozzi 1997: Maria Sofia Corradini Bozzi (ed.), *Ricettari medico-farmaceutici medievale nella Francia meridionale,* volume 1 (Accademia Toscana di Scienze e Lettere "La colombaria", Studi 159). Florence: Leo S. Olschki, 1997.

Corradini Bozzi 2023: Maria Sofia Corradini Bozzi (ed.), *Ricettari di Auch* in Repertorio informatizzato dell'antica letteratura trobadorica e occitana (RIALTO). See "Testi pratici", subheader "Letteratura medico-farmaceutica" http://www.rialto.unina.it/testipratici/medicafarm/Auch.htm.

Coutumes anciennes de Montpellier: Jeffrey Widmayer (ed.), *Coutumes anciennes de Montpellier (Ms. Montpellier, Bibliothèque de Médecine de l'Université de Montpellier H119).* PhD dissertation, University of North Carolina, Chapel Hill, 2004. Data from TMAO http://tmao.aieo.org/CoutMontpW_show2.

Daude de Prades, *Dels auzels cassadors*: Alexander H. Schultz (ed.), *The Romance of Daude de Prades called* Dels Auzels Cassadors. Columbus: Ohio State University Press, 1945.

Daurel e Beton: Charmaine Lee (ed.), *Daurel e Beton* (Biblioteca medievale 19). Parma and Roma: Pratiche Editrice and Carocci, 1991.

De modus viaticorum preparandorum et salsarum: Paris, Bibliothèque nationale de France, fonds latin, 8435, ff. 24 verso–28 verso.

"Eine Diätregel für einen Bischof": Karl Sudhoff (ed.), "Eine Diätregel für einen Bischof, aufgestellt von vier Professoren von Montpellier in der Mitte des 14. Jahrhunderts," *Archiv für Geschichte der Medizin* 14 (1923–1924), pp. 184–186.

Elucidari: Cyril P. Hershon, Peter T. Ricketts et al. (eds), *Elucidari de las proprietatz de totas res naturals* (Cahiers de Carrefour Ventadour). Moustier Ventadour: Carrefour Ventadour, 2018.

Flamenca: Ulrich Gschwind (ed.), *Le Roman de Flamenca, nouvelle occitane du 13e siècle.* Berne: Francke, 1976.

"A Fourteenth-century Provençal Receptary": Frances E. S. Norwood (ed.), "A Fourteenth-century Provençal Receptary," *Proceedings of the Leeds Philosophical and Literary Society* 17 (1978), pp. 171–254.

Girart de Roussillon: Micheline Combarieu du Grès and Gérard Gouiran (eds and transl.), *La Chanson de Girart de Roussillon* (Lettres gothiques). Paris: Librairie Générale Française, 1993.

Girart de Roussillon: W[inifred] Mary Hackett (ed.), *Girart de Roussillon, chanson de geste,* 3 volumes (Société des anciens textes français). Paris: A. &. J. Picard, 1953–1955.

Groenke 1986: Frank-Dieter Groenke (ed.), *Die frühmittelalterlichen lateinischen Monatskalendarien. Text – Übersetzung – Kommentar.* Dissertation for the degree of D. M. D. Berlin: Institut für Geschichte der Medizin der Freien Universität Berlin, 1986.

Guilhem Molinier, *Las Flors del Gay Saber*: Joseph Anglade (ed.), *Las Flors del Gay Saber* (Institut d'Estudis Catalans, Secció Filològica, Memòries I, 2). Barcelona: Institut d'Estudis Catalans, 1926.

Guilhem Molinier, *Las Leys d'amors*: Beatrice Fedi (ed.), *Las Leys d'amors, redazione lunga in prosa* (Archivio romanzo 35). Florence: Edizione del Galuzzo per la Fondazione Ezio Franceschini, 2019.

Guiot de Provins, La Bible: John Orr (ed.), *Les Œuvres de Guiot de Provins, poète lyrique et satirique.* Manchester: Imprimerie de l'Université, 1915.

Helmreich 1889: Georg Helmreich (ed.), *Marcelli, De Medicamentis* (Bibliotheca Scriptorum Graecorum et Romanorum Teubneriana). Leipzig: B. G. Teubner, 1889.

Henricus de Mondeville, *La Notomia*: Laure Combes (ed.), La *Notomia de Anric de Mondavilla*: traduction occitane de l'*Anatomia* d'Henri de Mondeville. Thesis for diplôme d'archiviste paléographique, École des Chartes, Paris. *Positions des thèses de l'École des Chartes* (2008) online.

Hunt 1990: Tony Hunt (ed.), *Popular Medicine in Thirteenth-Century England: Introduction and Texts.* Cambridge: D. S. Brewer, 1990.

Hunt 1994: Tony Hunt (ed.*), Anglo-Norman Medicine*. 2 volumes. Woodbridge, Suffolk: D. S. Brewer, 1994.

Hunt 2011: Tony Hunt (ed.), *Old French Medical Texts* (Textes littéraires du Moyen Âge 18). Paris: Classiques Garnier, 2011.

Ibn Butlan, *Taqwim al-Sihha*: Hosam Elkhadem (ed.), *Le Taqwim al-Sihha (Tacuini sanitatis) d'Ibn Butlan: un traité médical du XIe siècle* (Académie royale de Belgique, classe des lettres, fonds René Draguet 7). Louvain: Peeters, 1990.

Jaufre: Charmaine Lee (ed.), *Jaufre* (Biblioteca medievale 105). Rome: Carocci, 2006.

Liber de minucione et medicina secundum variationes lunarum: Paris, Bibliothèque nationale de France, fonds latin 10448, ff. 122 recto–122 verso.

Libre de Sent Soví: Rudolf Grewe (ed.), *Libre de Sent Soví. Llibre de totes maneres de potatges de menjar*, rev. Amadeu-J. Soberanas and Joan Santanach, 2nd ed. (Els nostres clàssics B 22). Barcelona: Barcino, 2004.

Livre de Sydrac: Paris, Bibliothèque nationale de France, fonds français 1158, ff. 1 recto–139 verso.

Livre Rouge des consuls de Montauban: Montauban, Archives départementales de Tarn-et-Garonne, E(03) 0121-AA 1, ff. 1 recto–115 recto.

Matfre Ermengaud, *Breviari 2*: Peter T. Ricketts (ed.), *Le* Breviari d'amor *de Matfre Ermengaud,* tome 2 *(1–8880)* (Association internationale d'études occitanes 4). London: AIEO and Westfield College, 1989.

Matfre Ermengaud, *Breviari 4*: Peter T. Ricketts, with Cyril P. Hershon (eds), *Le* Breviari d'amor *de Matfre Ermengaud,* tome 4 *(16783T–27252)* (Association internationale d'études occitanes 11). Turnhout: Brepols, 2004.

Matfre Ermengaud, *Breviari 5*: Peter T. Ricketts (ed.), *Le* Breviari d'amor *de Matfre Ermengaud,* tome 5 *(27252T–34597)*. Leiden: E. J. Brill, 1976.

Medicina Ypocratis quid usitare debeatur per singulos menses: London, British Library, Harley 3271, ff. 122 verso–123 recto.

Mesnagier de Paris, le: Georgina E. Brereton and Janet M. Ferrier (eds), Karin Ueltschi (transl.), *Le Mesnagier de Paris* (Lettres gothiques). Paris: Livre de poche, 1994.

Meyer 1903: Paul Meyer (ed.), "Recettes médicales en provençal d'après le ms. R. 14.30 de Trinity College (Cambridge)," *Romania* 32 (1903), pp. 268–299 and 472–474.

Meyer 1909: Paul Meyer (ed.), *Documents linguistiques du Midi de la France: Ain, Basses-Alpes, Hautes-Alpes, Alpes-Maritimes*. Paris: Honoré Champion, 1909.

Mystères provençaux, les: Alfred Jeanroy and Henri Teulié (eds), *Les Mystères provençaux du quinzième siècle* (Bibliothèque méridionale 3). Toulouse: Privat, 1893.

"Names of the Mother of God, the": Paris, Bibliothèque nationale de France, nouvelle acquisition française 4232, ff. 9 recto–13 verso.

Paden 2016: William D. Paden (ed.), *Two Medieval Occitan Toll Registers from Tarascon* (Medieval Academy Books 115). Toronto: University of Toronto Press, 2016.

Peire Raimon, "Poésies du troubadour Peire Raimon de Toulouse": Joseph Anglade (ed.), "Poésies du troubadour Peire Raimon de Toulouse," *Annales du Midi: revue archéologique, philologique de la France méridionale* 31 (1919), pp. 157–189.

Peire Raimon, "Poésies du troubadour Peire Raimon de Toulouse, suite": Joseph Anglade (ed.), "Poésies du troubadour Peire Raimon de Toulouse, suite et fin," *Annales du Midi: revue archéologique, philologique de la France méridionale* 32 (1920), pp. 257–304.

Francés Pellos, *Compendion de l'abaco*: Robert Lafont (ed.), *Francés Pellos, Compendion de l'abaco. Texte établi d'après l'édition de 1492*. Montpellier: Université de Montpellier, 1967.

Petit Thalamus de Montpellier: Ferdinand Pegat and Eugène Thomas (eds), *Le petit Thalamus de Montpellier* (Publications de la Société archéologique de Montpellier 2, 4–5, 8–9, 11). Montpellier: Jean Martel, 1840.

Petrus Fagarola, "Advice from a Physician to His Sons": Lynn Thorndike (ed.), "Advice from a Physician to His Sons," *Speculum* 6 (1931), pp. 110–114.

Petrus Fagarola, "Translation of a Letter from a Physician of Valencia to His Two Sons Studying at Toulouse 1315 A.D.": Lynn Thorndike (transl.), "Translation of a Letter from a Physician of Valencia to His Two Sons Studying at Toulouse 1315 A.D.," *Annals of Medical History* 3 (1931), pp. 17–20.

Peyre de Serras, "Recettes pharmaceutiques": Florence, Biblioteca Medicea Laurenziana, Ashburnham 40, vol. 1: ff. 93 recto–100 recto; vol. 2: ff. 9 verso–20 verso and 24 verso–50 recto. See "A Fourteenth-century Provençal Receptary" (ed. Norwood 1978).

Provenzalische Diätetik: Hermann Suchier (ed.), *Provenzalische Diätetik auf Grund neuen Materials*. Halle: Max Niemeyer, 1894.

Raimon de Perelhos, *Voyage au Purgatoire de St Patrice*: Alfred Jeanroy and Alphonse Vignaux (eds), *Voyage au Purgatoire de St Patrice, Visions de Tindal et de St Paul* (Bibliothèque méridionale 8). Toulouse: Édouard Privat, 1903.

Raimon Feraut, *La Vida de sant Honorat*: Ingegärd Suwe (ed.), *La Vida de sant Honorat*. Uppsala: A.-B. Lundequistka Bokhandeln, 1943.

Ramon Vidal de Besalú, *Judici d'Amor*: Hugh Field (ed.), *Obra poètica II*. Barcelona: Curial, 1991.

Regola sanitaria salernitana: Fulvio Gherli (transl.), *Regola sanitaria salernitana, Regimen sanitatis Salernitanum*. Rome: SEDAC for the Ente provinciale per il turismo, 198–.

Rogerius of Salerno, *Chirurgia*: Basel, Universitätsbibliothek, D.II.11, ff. 139 recto–153 recto.

"Recettes médicales": Basel, Universitätsbibliothek, D.II.11, ff. 153 verso–154 recto.

Regimen XII mensium: Montpellier, Bibliothèque Universitaire Historique de Médecine, Université de Montpellier, H 185, ff. 160 recto–160 verso.

Roland occitan: Gérard Gouiran and Robert Lafont (eds and transl.), *Le Roland occitan: Roland à Saragosse, Ronsasvals* (Bibliothèque médiévale). Paris: Christian Bourgeois, 1991.

Roman d'Esther: Suzanne Thiolier and Marie-Françoise Notz-Grob (eds and transl.), *Nouvelles courtoises occitanes et françaises* (Lettres Gothiques). Paris: Livre de Poche, 1997.

Roman d'Arles: Hans-Christian Haupt (ed.), *Le "Roman d'Arles" dans la copie de Bertran Boysset (manuscrit Aix-en-Provence, Musée Paul Arbaud, M.O.63): études et édition* (Romanica Helvetica 121). Tübingen and Basel: A. Francke, 2003.

Rosari, lo: Antoine Calvet (ed.), *Le Rosier alchimique de Montpellier:* Lo Rosari *(XIVe siècle)* (Centre d'enseignement et de recherche d'Oc 9). Paris: Presses de l'Université de Paris-Sorbonne, 1997.

Santanach 2008: Joan Santanach (ed.), Robin Vogelzang (transl.), *The Book of Sent Soví: Medieval Recipes from Catalonia*. Barcelona and Woodbridge: Barcino-Tamesis, 2008.

Stuart 1979: Heather Stuart, "A Ninth-century Account of Diets and *Dies Aegyptiaci*," *Scriptorium* 33 (1979), pp. 237–244.

Sudhoff 1908: Karl Sudhoff (ed.), "Medizinische Monatsregeln für Aderlaß, Schröpfen, Baden, Arzneigebrauch und Auswahl der Speisen und Getränke aus einer Pariser Handschrift des 14. Jahrhunderts," *Archiv für Geschichte der Medizin* 2 (1908), pp. 136–139.

Thiolier-Méjean 1999: Suzanne Thiolier-Méjean, *L'Alchimie médiévale*: L'obratge dels philosophes; La soma; *et les manuscrits d'oïl* (Centre d'enseignement et de recherche d'oc 10). Paris: Presses de l'Université de Paris-Sorbonne, 1999.

"Traité des noms de la mère de Dieu": Paul Meyer (ed.), "Traité des noms de la mère de Dieu," published with *Daurel et Beton: Chanson de geste provençale* (Société d'anciens textes français). Paris: Firmin Didot, 1880, pp. c–cviii.

***Trésor manuscrit de l'ancien occitan* 2017–**: Dominique Billy (dir.), *Trésor manuscrit de l'ancien occitan*. Toulouse: Université Jean-Jaurès and CNRS, 2017–. http://tmao.aieo.org.

Uc Faidit, *Donatz proensals*: John H. Marshall (ed.), *The* Donatz proensals *of Uc Faidit* (University of Durham Publications). London: Oxford University Press, 1969.

"Vocabulaire provençal-latin": Alphonse Blanc (ed.), "Vocabulaire provençal-latin," *Revue des langues romanes* 35 (1891), pp. 29–87 and 319–320.

Secondary Literature

Adams 1913: Edward L. Adams, *Word-Formation in Provençal* (University of Michigan Studies, Humanistic Series 2). New York: The Macmillan Company, 1913.

Adamson 1995: Melitta Weiss Adamson, *Medieval Dietetics: Food and Drink in* Regimen Sanitatis *Literature from 800 to 1400* (German Studies in Canada 5). Frankfurt am Main: Peter Lang, 1995.

Adamson 2002: Melitta Weiss Adamson (ed.), *Regional Cuisines of Medieval Europe: A Book of Essays*. London: Routledge, 2002.

Adamson 2004: Melitta Weiss Adamson, *Food in Medieval Times* (Food through History). Westport, CT: Greenwood Press, 2004.

Appel 1920: Carl Appel (ed.), *Provenzalische Chrestomathie mit Abreiss der Formenlehre und Glossar.* 5th ed. Leipzig: O. R. Reisland, 1920.

Archives de littérature du Moyen Âge: Archives de littérature du Moyen Âge (ARLIMA). www.arlima.net/index.html.

Banegas López 2010: Ramón A. Banegas López, "Consumption of Meat in Western European Cities during the Late Middle Ages: A Comparative Study," *Food & History* 8 (2010), pp. 63–86.

Banegas López 2017: Ramón A. Banegas López, "Évolution de la consommation de viande de mouton en Catalogne au long du Moyen Âge: La construction d'un modèle alimentaire," in Bruno Laurioux and Marie-Pierre Horard (dir.), *Pour une histoire de la viande* (Tables des hommes). Tours: Presses universitaires François-Rabelais, 2017, pp. 151–163; cited from *Pour une histoire de la viande* [on line]. Tours: Presses universitaires François-Rabelais, 2017 (generated 2 May 2023). Available at: http://books.openedition.org/pufr/25782.

Barbaud 1986: Jean Barbaud, "Les boissons dans les calendriers diététiques," *Bulletin de liaison de l'Assocation des Amis du Musée de la Pharmacie de Montpellier* 11 (1986), pp. 15–26.

Barbaud 1988: Jean Barbaud, "Les anciens calendriers diététiques," *Revue d'histoire de la pharmacie* 76 (1988), pp. 328–342.

Barbaud and Gillon 1988: Jean Barbaud and Pierre Gillon, "Cinq calendriers diététiques provenant de manuscrits de la Bibliothèque nationale de Paris," in Hans Schadewaldt (ed.), *Actes, XXXe Congrès international d'histoire de la médecine, Düsseldorf 31.VIII–5.IX 1986*. Düsseldorf: Organisationskomitee des XXX. Internationalen Kongresses, 1988, pp. 179–187.

Bec 1977–1978: Pierre Bec, *La lyrique française au Moyen Âge (XIIe–XIIIe siècles). Contribution à une typologie des genres poétiques médiévales*. 2 volumes. Paris: A. & J. Picard, 1977–1978.

Bec 1987: Pierre Bec (ed.), *Anthologie de la prose occitane du Moyen Âge (XIIe–XVe siècle)*. Valdarias: Vent Terral, 1987.

Beccaria 1956: Augusto Beccaria, *I codici di medicina del periodo presalernitano (secoli IX, X e XI)*. Rome: Storia e letteratura, 1956.

Bedel 1996: Christian-Pierre Bedel (ed.), *Sévérac, Buzeins, Lapanouse, Lavernhe, Recoules-Prévinquières (Severac, Busens, La Panosa, Recolas-Previnquièiras, La Vèrnha)* (Al canton). Rodez: Mission départementale de la culture, 1996.

Bénézet 1999: Jean-Pierre Bénézet, *Pharmacie et médicament en Méditerranée occidentale (XIIIe–XVIe siècles)* (Sciences, techniques et civilisations du Moyen Âge à l'aube des Lumières 3). Paris: Honoré Champion, 1999.

Bérier 1988: François Bérier, "XIV. La Traduction en français," in Daniel Poirion (dir.), Armin Biermann, Dagmar Tillmann-Bartylla (eds), *La Littérature française aux XIVe et XVe siècles*, in *Grundriß der romanischen Literaturen des Mittelalters*, vol 8. Heidelberg: Carl Winter Universitätsverlag, 1988, t. 1, pp. 219–265.

Binz 1908: Gustav Binz, *Handwritten Description of D. II.11. 23 March 1908*. Available at https://swisscollections.ch/Record/991170506236105501/Holdings#tabnav.

Bizzari 2014: Hugo Bizzari, "E. Le Secretum secretorum," in Catherine Gaullier-Bougassas (ed.), *La fascination pour Alexandre le Grand dans les littératures européennes (Xe–XVIe siècle)*, volume 1. *Réinventions d'un mythe* (Alexander Redivivus 6). Turnhout: Brepols, 2014, pp. 44–45.

Black 2020: Winston Black (ed.), *Medicine and Healing in the Premodern West: A History in Documents* (Broadview Sources Series). Peterborough: Broadview Press, 2020.

Bondurand 1895: Édouard Bondurand (transl.), "Une diététique provençale," *Revue du Midi* 18 (1895), pp. 191–207.

Borland 2022: Jennifer Regan Borland, *Visualizing Household Health: Medieval Women, Art, and Knowledge in the* Régime du Corps. University Park, PA: Pennsylvania State University Press, 2022.

Bos et al. 2021: Gerrit Bos, Sandra Hajek, and Guido Mensching, "A Fragment of Abraham Avigdor's Translation of Gerard de Solo's *Practica* from the Cairo Genizah: Edition and Analysis with Special Regard to the Occitan Elements." *Aleph* 21.2 (2021), pp. 309–357.

Braekman 1986: Willy Louis Braekman, "Queen Isabel's Dietary and Its Context," in W. L. Braekman (ed.), *Studies on Alchemy, Diet, Medecine and Prognostication in Middle English*. Brussels: Omirel, UFSAL, 1986, pp. 43–83.

Brinckmann 1914: Johannes Brinckmann, *Die apokryphen Gesundheitsregeln des Aristoteles für Alexander den Grossen in der Übersetzung des Johann von Toledo*. Inaugural-Dissertation. Institut für Geschichte der Medizin zu Leipzig. Leipzig: Metzger & Wittig, 1914.

Bruegel and Laurioux 2002: Martin Bruegel and Bruno Laurioux (eds), *Histoire et identités alimentaires en Europe*. Paris: Hachette, 2002.

Brunel 1935: Clovis Brunel, *Bibliographie des manuscrits littéraires en ancien provençal* (Société de publications romanes et françaises 13) Paris: Librairie E. Droz, 1935.

Buridant 1983–1984: Claude Buridant, "Séminaire de Philologie romane, Université de Strasbourg, 1983–1984" (series of university lectures).

Calames: See Catalogue en ligne des archives et des manuscrits de l'enseignement supérieur, below.

Cappelli 1998: Adriano Cappelli, *Cronologia: Cronografia e calendario perpetuo*, 7th ed. (Manuali Hoepli). Milano: Ulrico Hoepli, 1998.

Carré 2019: Antònia Carré, "'De les coses que el cos nodreixen': Els aliments al *Regiment de sanitat* d'Arnau de Vilanova," in Jordi Salas-Salvadó, Montserrat Barbany, and Abel Mariné (eds), *Alimentació i nutrició a Catalunya: Industrials, productors i científics*. Barcelona: Associació catalana de ciències de l'alimentació de l'Institut d'Estudis Catalans, 2019, pp. 171–185.

Catalo et al. 1995: Jean Catalo, Isabelle Rodet-Belarbi, and Yves Lignereux, "Déchets de boucherie et alimentation au XIVe siècle à l'Hôpital du Pas à Rodez," *Archéologie du Midi médiéval* 13 (1995), pp. 187–195.

Catalogue en ligne des archives et des manuscrits de l'enseignement supérieur: Catalogue en ligne des archives et des manuscrits de l'enseignement supérieur online. http://www.calames.abes.fr/pub/.

Chambon 1994: Jean-Pierre Chambon, Review of Ingrid Arthur (ed.), *Miracles que Dieus ha mostratz per sant Frances apres la sua fin: Version occitane de la Legenda Maior Sancti Francisci, Miracula de saint Bonaventure, Revue de linguistique romane* 58 (1994), pp. 565–572.

Cifuentes i Comamala 1999: Lluís Cifuentes i Comamala, "Vernacularization as an Intellectual and Social Bridge: The Catalan Translations of Teodorico's *Chirurgia* and of Arnau de Vilanova's *Regimen sanitatis*." *Early Science and Medicine* 4 (1999), pp. 127–148.

Cifuentes i Comamala 2004: Lluís Cifuentes i Comamala, "Université et vernacularisation au bas Moyen Âge: Montpellier et les traductions catalanes médiévales de traités de médecine," in Daniel Le Blévec and Thomas Granier (eds), *L'Université de médecine de Montpellier et son rayonnement (XIII–XVe siècles): Actes du colloque international de Montpellier (Université Paul Valéry, Montpellier III, 17–19 mai, 2001)* (Collection De diversis artibus 71 [n.s. 34]). Turnhout: Brepols, 2004, pp. 273–290.

Cifuentes i Comamala 2016: Lluís Cifuentes i Comamala, "El receptari mèdic baixmedieval i renaixentista: un gènere vernacle," in Lola Badia, Lluís Cifuentes, Sadurní Martí, and Josep Pujol (eds), *Els manuscrits, el saber i les lletres a la Corona d'Aragó, 1250–1500*. Barcelona: Publicacions de l'Abadia de Montserrat, 2016, pp. 103–160.

Combes 1990: Jean Combes, "Un groupe d'hommes d'affaires montpelliérains à la fin du XIVe siècle et au commencement du XVe," in Jean Combes, *Montpellier et le Languedoc au Moyen Âge* (Mémoires de la Société archéologique de Montpellier 20). Montpellier: Société archéologique de Montpellier, 1990, pp. 85–120.

Comité d'études gascon de la langue occitane 1998: Comité d'études gascon de la langue occitane (Hautes-Pyrénées). *Atau que's ditz!: Dictionnaire français-occitan, gascon des Hautes-Pyrénées*. Tarbes: Comitat dera lenga, Conseil général des Hautes-Pyrénées, 1998.

***Concordance de l'occitan médiéval* 2005**: Peter T. Ricketts et al. (eds), *Concordance de l'occitan médiéval*, 2 tranches. Turnhout: Brepols, 2005 [CD-Rom].

Corradini Bozzi 2012: Maria Sofia Corradini Bozzi, "La letteratura medica medievale in lingua d'oc fra tradizione antica e rinascimento europeo," in Anna Alberni (ed.), *El saber i les llengües vernacles a l'època del Llull i Eiximenis: estudis ICREA sobre vernacularizació* (Textos i Estudis de Cultura Catalana). Barcelona: Publicacions de l'Abadia de Montserrat, 2012, pp. 147–170.

Coulet 1988: Noël Coulet, *Aix-en-Provence: espace et relations d'une capitale (milieu XIVe s.–milieu XVe s.).* Aix-en-Provence: Université de Provence, 1988.

Cummins 1976: Cummins, Patricia Willet. *A Critical Edition of "Le Regime tresutile et tresproufitable pour conserver et garder la santé du corps humain" with the Commentary of Arnoul de Villeneuve, Corrected by the Docteurs regens of Montpellier 1480; Lyon 1491* (North Carolina Studies in the Romance Languages and Literatures 177). Chapel Hill: University of North Carolina Press, 1976.

Daniel 1914: Jean Daniel, *Dictionnaire périgourdin.* 1ère partie: *Dictionnaire français-périgourdin avec supplement.* Périgueux: Ribes, 1914.

Demaitre 1980: Luke E. Demaitre, *Doctor Bernard de Gordon: Professor and Practitioner* (Studies and Texts 51). Toronto: Pontifical Institute of Mediaeval Studies, 1980.

Demaitre 2004: Luke E. Demaitre, "Bernard de Gordon et son influence sur la pensée médicale aux XIVe et XVe siècles," in Daniel Le Blévec and Thomas Granier (eds), *L'Université de médecine de Montpellier et son rayonnement (XIII–XVe siècles: Actes du colloque international de Montpellier (Université Paul Valéry, Montpellier III, 17–19 mai, 2001)* (Collection De diversis artibus 71 [n.s. 34]). Turnhout: Brepols, 2004, pp. 103–131.

Dictionary of Old Occitan Medico-Botanical Terminology (DiTMAO): *Dictionary of Old Occitan Medico-Botanical Terminology (DiTMAO).* https://www.uni-goettingen.de/en/487498.html.

Dictionnaire de l'occitan médiéval: Dictionnaire de l'occitan médiéval online. www.dom-en-ligne.de/.

Drabkin 1942: Miriam Drabkin, "Select Pages from Mediaeval Medical Manuscripts." *Bulletin of the History of Medicine* 11 (1942), pp. 409–439.

Ducos 1999: Joëlle Ducos, "L'écrit scientifique au Moyen Âge – langue d'oc et langue d'oïl." *Garona, Cahier du CECAES* 15 (1999), pp. 55–71.

Dumas 2015: Geneviève Dumas, *Santé et société à Montpellier à la fin du Moyen Âge* (The Medieval Mediterranean 102). Leiden: Brill, 2015.

Faure 2009: Andrieu Faure, *Assai de diccionari francés-aupenc d'òc.* Gap: Rescontre Gapian, 2009.

Fery-Hue 1992: Françoise Fery-Hue, "Secret des secrets," in Robert Bossuat et al. (eds), rev. Geneviève Hasenohr and Michel Zink (eds), *Dictionnaire des lettres françaises: Le Moyen Âge*, Paris: Fayard 1992, pp. 1366–1370.

Fischer 2000a: Klaus-Dietrich Fischer, *Bibliographie des textes médicaux latins: Antiquité et haut Moyen Âge*, premier supplément, 1986–1999 (Centre Jean-Palerne, Mémoires 19). Saint-Étienne: Publications de l'Université de Saint-Étienne, 2000.

Fischer 2000b: Klaus-Dietrich Fischer, "The 'Isagoge' of Pseudo-Soranus: An Analysis of the Contents of a Medieval Introduction to the Art of Medicine," *Medizinhistorisches Journal* 35 (2000), pp. 3–30.

Fischer 2015: Klaus-Dietrich Fischer, "Gesund durchs Jahr mit Dr. Hippokrates – Monat für Monat!" in Brooke Holmes and Klaus-Dietrich Fischer (eds) with the assistance of Emilio Capettini, *The Frontiers of Ancient Science. Essays in Honor of Heinrich von Staden* (Beiträge zur Altertumskunde 338). Berlin: De Gruyter, 2015, pp. 111–137.

Forestié 1884: Édouard Forestié, *Nos pères à table. Étude sur l'alimentation des Montalbanais aux XIIe, XIVe et XVIe siècles.* Montauban: Imprimerie Forestié, 1884.

Foulet and Speer 1979: Alfred Foulet and Mary Blakely Speer, *On Editing Old French Texts.* Lawrence: The Regents Press of Kansas, 1979.

Galderisi 2011: Claudio Galderisi (dir.), *Translations médiévales: Cinq siècles de traductions en français au Moyen Âge (XIe–XVe siècles), étude et répertoire.* 2 volumes in 3 tomes. Turnhout: Brepols, 2011.

Gallica: Gallica, Bibliothèque nationale de France. https://gallica.bnf.fr/accueil/en/content/accueil-en?mode=desktop.

Gamer, Helena M. 1954: "The Earliest Evidence of Chess in Western Literature," *Speculum* 29.4 (October 1954), pp. 734–750.

Garrison 1929: Fielding H. Garrison, *An Introduction to the History of Medicine*, 4th ed. Philadelphia and London: Saunders, 1929.

Gaster 1907–1908: M. Gaster, "Introduction to a Hebrew Version of the *Secret of Secrets*," *Journal of the Royal Asiatic Society* 1907, pp. 879–913; part 1, 1908, pp. 111–162; part 2, 1908, pp. 1065–1084.

Gaullier-Bougassas 2014: Catherine Gaullier-Bougassas, "D. Les textes didactiques" in Catherine Gaullier-Bougassas (ed.), *La fascination pour Alexandre le Grand dans les littératures européennes (Xe–XVIe siècle)*, volume 1. *Réinventions d'un mythe* (Alexander Redivivus 5). Turnhout: Brepols, 2014, p. 1: 54.

Gaullier-Bougassas et al. 2015a: Catherine Gaullier-Bougassas, Margaret Bridges, and Jean-Yves Tilliette, "Cheminements culturels et métamorphoses d'un texte aussi célèbre qu'énigmatique," in Catherine Gaullier-Bougassas, Margaret Bridges, and Jean-Yves Tilliette (eds), *Trajectoires européennes du* Secretum secretorum *du Pseudo-Aristote (XIIIe–XVIe siècle)* (Alexander Redivivus 6). Turnhout: Brepols, 2015, pp. 5–25.

Gaullier-Bougassas et al. 2015b: Catherine Gaullier-Bougassas, Margaret Bridges, and Jean-Yves Tilliette (eds), *Trajectoires européennes du* Secretum secretorum *du Pseudo-Aristote (XIIIe–XVIe siècle)* (Alexander Redivivus 6). Turnhout: Brepols, 2015.

Gérard-Zai 2009: Marie-Claire Gérard-Zai, "Les réceptaires occitans de la fin du Moyen Âge," in Guy Latry (ed.), *La voix occitane. Actes du VIIIe congrès de l'Association internationale d'études occitanes, Bordeaux, 12–17 octobre 2005*, volume 1. Pessac: Presses universitaires de Bordeaux, 2009, pp. 527–534.

Gil-Sotres et al. 1996: Pedro Gil-Sotres, with Juan Antonio Paniagua and Luis García-Ballester, "Estudi introductory," in Luis García-Ballester and Michael R. McVaugh (eds), Arnaldi de Villanova, *Opera Medica Omnia* X.1, *Regimen sanitatis ad regem Aragonum*, Barcelona: Seminarium historiae scientiae Barchinone, 1996, pp. 17–421 and 471–933.

Gougaud 1989: Henri Gougaud (transl.), *Chanson de la croisade albigeoise* (Lettres gothiques). Paris: Livre de poche, 1989.

Gran Diccionari de la llengua catalana: *Gran Diccionari de la llengua catalana*. https://www.diccionari.cat/gran-diccionari-de-la-llengua-catalana#.

Green 2019: Monica H. Green, "Recovering 'Ancient' Gynaecology: The Humanist Rediscovery of the Eleventh-Century Gynaecological Corpus," in Outi Merisalo, Miika Kuha, and Susanna Niiranen (eds), *Transmission of Knowledge in the Late Middle Ages and the Renaissance* (Bibliologia 53). Turnhout: Brepols, 2019, pp. 45–54.

Green 2020: Monica H. Green, "Medicine in France and England in the Long Twelfth Century: Inheritors and Creators of European Medicine," in Charlotte Denoël and Francesco Siri (eds), *France et Angleterre: manuscrits médiévaux entre 700 et 1200*. Turnhout: Brepols, 2020, pp. 363–388.

Guénoun 1999: Anne-Sylvie Guénoun, "Le médicament chez Gérard de Solo, médecin montpelliérain du XIVe siècle," *Revue d'histoire de la pharmacie* 324 (1999), pp. 465–474.

Halleux and Opsomer 2005: Robert Halleux and Carmélia Opsomer, "L'insaisissable médecine populaire," in Pierre Nobel (ed.), *La Transmission des savoirs au Moyen Âge et à la Renaissance*, volume 1: *Du XIIe au XVe siècle* (Littéraires). Besançon: Presses universitaires de Franche-Comté, 2005, pp. 331–348.

Hershon 2023: Cyril P. Hershon, *Le* Breviari d'amor *de Matfre Ermengaud*, 1: *Introduction*, rev. Dominique Billy (Publications de l'Association internationale d'études occitanes 15). Turnhout: Brepols, 2023.

Hofmeister-Winter et al. 2014: Andrea Hofmeister-Winter, Helmut W. Klug, and Karin Kranich (eds), *Der Koch ist der bessere Artz: Zum Verhältnis von Diätetik und Kulinarik im Mittelalter und in der Frühen Neuzeit. Fachtagung im Rahmen des Tages der Geisteswissenschaften 2013 an der Karl-Franzens-Universität Graz, 20.6–22. 6. 2013* (Mediävistik zwischen Forschung, Lehre und Öffentlichkeit 8). Frankfurt-am-Main: Peter Lang, 2014.

Hunt 1999: Tony Hunt, "Old French Translation of Medical Texts," *Forum for Modern Language Studies* 35 (1999), pp. 350–357.

Hyman and Hyman 1983: Philip Hyman and Mary Hyman, "Les cuisines régionales à travers des livres de recettes," *Dix-huitième siècle* 15 (1983), pp. 65–74.

Institut d'Estudis Catalans, *Diccionari*: Institut d'Estudis Catalans, *Diccionari català-valencià-balear.* https://dcvb.iec.cat/.

Jacquart 1998: Danielle Jacquart, "Medical Scholasticism," in Mirko Grmek and Bernardino Fantini (eds), Antony Shugaar (transl.), *Western Medical Thought from Antiquity to the Middle Ages*. Cambridge, MA: Harvard University Press, 1998, pp. 197–240, 380–387. Translation of "La scolastica medica," originally published in *Storia del pensiero medico occidentale*, volume 1. Bari: Laterza, 1993, pp. 261–321.

Jonas: Jonas, Répertoire des textes et des manuscrits médiévaux d'oc et d'oïl. https://jonas.irht.cnrs.fr.

Jud 1934: Jakob Jud, "Sur l'histoire de la terminologie ecclésiastique de la France et de l'Italie (avec 7 cartes)," *Revue de linguistique romane* 10 (1934), pp. 1–62.

Jud 1939: Jakob Jud, "Observations sur le lexique du roman de *Flamenca*," in *Mélanges A. Duraffour: Hommage offert par ses amis et ses élèves 4 juin 1939*. Paris and Zurich: E. Droz and Max Niehans, 1939, pp. 204–210.

Kallinich and Figala 1972: Günter Kallinich and Karin Figala, "Das 'Regimen sanitatis' des Arnold von Bamberg," *Sudhoffs Archiv* 56.1 (1972), pp. 44–60.

Katara 1966: Pekka Katara, *Das französische Lehngut in mittelniederdeutschen Denkmälern von 1300 bis 1600* (Mémoires de la Société néophilologique de Helsinki 30). Helsinki: Société néophilologique, 1966.

Kibre 1945: Pearl Kibre, "Hippocratic Writings in the Middle Ages," *Bulletin of the History of Medicine* 18 (1945), pp. 371–412.

Lafont 1983: Robert Lafont, "Lenga d'oc e umanisme scientific," *Amiras / Repères* 5 (1983), pp. 33–36 + 1 ill.

Lambert 1989a: Carole Lambert, "La cuisine française au bas Moyen Âge: pays d'oïl vs pays d'oc," in Jean Peltre and Claude Thouvenot (eds), *Alimentation et régions. Actes du colloque Cuisines, régimes alimentaires, espaces régionaux, Nancy, 24–27 septembre 1987.* Nancy: Presses universitaires de Nancy, 1989, pp. 375–385.

Lambert 1989b: Carole Lambert (ed.), "Trois réceptaires culinaires médiévaux: *Les Enseingnemenz*, les *Doctrine* et le *Modus*. Édition critique et glossaire détaillé." Doctoral dissertation, Université de Montréal, 1989.

Lambert 1992: Carole Lambert (ed.), *Du manuscrit à la table. Essais sur la cuisine au Moyen Âge.* Montréal: Presses de l'université de Montréal, 1992.

Lambert 1997–1998: Carole Lambert, "La cuisine occitane à partir d'un réceptaire culinaire languedocien et de sources annexes," *Archéologie du midi médiéval* 15 and 16 (1997–1998), pp. 295–305.

Lambert 2002: Carole Lambert, "Medieval France. B. The South," in Melitta Weiss Adamson (ed.), *Regional Cuisines of Medieval Europe: A Book of Essays*. London: Routledge, 2002, pp. 67–84.

Langlois 1889: Ernest Langlois, *Notices des manuscrits français et provençaux de Rome antérieurs au XVIe siècle*. Paris: Imprimerie nationale (from *Notices et extraits des manuscrits de la Bibliothèque nationale et autres bibliothèques* 33, 2e partie), 1889.

Lartigaut 2000: Jean Lartigaut, *Le Quercy après la guerre de Cent ans (vers 1440–vers 1500): aux origines du Quercy actuel.* Cahors: Éditions Quercy-Recherche, 2000; reprint and expanded version of *Les campagnes du Quercy après la guerre de Cent ans.* Toulouse: Association des publications de l'Université de Toulouse-le-Mirail, 1978.

Laurioux 1983: Bruno Laurioux, "De l'usage des épices dans l'alimentation médiévale," *Médiévales* 5 (1983), pp. 15–31.

Laurioux 2004: Bruno Laurioux, "'Il n'est bon bec qu'à Paris': la naissance d'une capitale gastronomique à la fin du Moyen Âge," in Claude Gauvard and Jean-Louis Robert (eds), *Être Parisien. Actes du* *colloque organisé par l'École doctorale d'histoire de l'Université Paris I Panthéon-Sorbonne et la Fédération des Sociétés historiques et archéologiques de Paris-Île-de-France (26–28 septembre 2002)* (Mémoires de la Fédération des Sociétés historiques et archéologiques de Paris et de l'Île-de-France 55, Homme et société 29). Paris: Publications de la Sorbonne, 2004, pp. 209–229.

Laurioux 2005: Bruno Laurioux, *Une histoire culinaire du Moyen Âge* (Sciences, techniques et civilisations du Moyen Âge à l'aube des Lumières 8). Paris: Honoré Champion, 2005.

Laurioux 2011: Bruno Laurioux, *Écrits et images de la gastronomie médiévale* (Conférences Léopold Delisle). Paris: Bibliothèque nationale de France, 2011.

Le Bret 1841: Henri Le Bret, *Histoire de Montauban*, revised edition by l'Abbé Marcellin and Gabriel Ruck, 2 volumes. Montauban: Rethoré, 1841; originally published as *Histoire de la ville de Montauban*, 2 volumes. Montauban: S. Dubois, 1668.

Levy 1894: Emil Levy, Review of Suchier, *Diététique provençale*, *Literaturblatt für germanische und romanische Philologie* 15 (1894), pp. 402–404.

Levy 1894–1924: Emil Levy, *Provenzalisches Supplement-Wörterbuch. Berichtigungen und Ergänzungen zu Raynouards* Lexique roman, 8 volumes. Leipzig: Reisland, 1894–1924, reprint Hildesheim: Olms, 1973.

Lignereux 2011: Yves Lignereux, "Nourritures carnées au Moyen-Âge. Situation dans les *castra* entre Pyrénées et Languedoc (XIIIe–XVe s.), seconde partie," *Bulletin du Centre d'étude d'histoire de la médecine (CEHM)* 77 (2011), pp. 27–54.

Matterer 2001: James Matterer (transl.), *A Boke of Gode Cooker – Regimen Sanitatis Salernitanum*. 2001. http://www.godecookery.com/regimen/regimen.htm.

McVaugh 1976: Michael McVaugh, "Two Montpellier Recipe Collections," *Manuscripta* 20 (1976), pp. 175–180.

McVaugh 2017: Michael McVaugh, "In a Montpellier Classroom," in Gideon Manning and Cynthia Klestinec (eds), *Professors, Physicians, and Practices in the History of Medicine: Essays in Honor of Nancy Siraisi* (Archimedes 50). Cham, Switzerland: Springer Verlag, 2017, pp. 57–76.

Mehl 2010: Jean-Michel Mehl, *Des jeux et des hommes dans la société médiévale* (Nouvelle Bibliothèque du Moyen Âge 97). Paris: Honoré Champion, 2010.

Mercadièr and Chapduelh 2023: Gilabert Mercadièr (dir.) and Danís Chapduelh (lexicographer), *Diccionari francés-occitan lemosin*. https://dicodoc.eu/oc/diccionaris.

Mistral 1878: Frédéric Mistral (ed.), *Lou Tresor dóu Felibrige ou Dictionnaire provençal-français*, 2 volumes. Aix-en-Provence and Avignon: veuve Remondet-Aubin and Roumanille, 1878, reprint Raphèle-les-Arles: Culture provençale et méridionale, Marcel Petit, 1979.

Monfrin 1982: Jacques Monfrin, "La place du *Secret des Secrets* dans la littérature française médiévale," in William F. Ryan and Charles B. Schmitt (eds), *Pseudo-Aristotle, The Secret of Secrets. Sources and Influences* (Warburg Institute Surveys 9). London: Warburg Institute, University of London, 1982, pp. 73–113.

Monson 1981: Don Alfred Monson, *Les "Ensenhamens" occitans: Essai de définition et de délimitation du genre*. Paris: Klincksieck, 1981.

Mourthé 2018: Claude Mourthé (transl.), *Chanson de la croisade albigeoise*. Paris: Belles lettres, 2018.

Mouynès 1871–1879: Germain Mouynès, *Ville de Narbonne. Inventaire des Archives communales antérieures à 1790*, 6 volumes. Narbonne: E. Caillard, 1871–1879.

Murray 1941: Harold J. R. Murray, "The Medieval Game of Tables," *Medium Aevum* 10 (1941), pp. 57–69.

Nicoud 2004: Marilyn Nicoud, "Les régimes de santé de l'aire montpelliéraine: affirmation et renouveau de l'*ars diaetae* au XIVe siècle," in Daniel Le Blévec and Thomas Granier (eds), *L'Université de Médecine de Montpellier et son rayonnement (XIIIe–XVe siècles): Actes du colloque international de Montpellier organisé par le Centre historique de recherches et d'études médiévales sur la Méditerranée occidentale (Université Paul Valéry – Montpellier III), 17–19 mai 2001* (Collection De diversis artibus 71 [n.s. 34]). Turnhout: Brepols, 2004, pp. 233–251.

Nicoud 2007: Marilyn Nicoud, *Les Régimes de santé au Moyen Âge: Naissance et diffiusion d'une écriture médicale (XIIIe–XVe siècle)*, 2 volumes (Bibliothèque des Écoles françaises d'Athènes et de Rome 333). Rome: École française de Rome, 2007.

Nicoud 2014: Marilyn Nicoud, *Le Prince et les médecins. Pensée et pratiques médicales à Milan (1402–1476)* (Collection de l'École française de Rome 488). Rome: École française de Rome, 2014.

Opsomer 1991: Carmélia Opsomer, *L'art de vivre en santé. Images et recettes du Moyen Âge, le* Tacuinum Sanitatis *(manuscrit 1041) de la Bibliothèque de l'Université de Liège*. Liège: Éditions du Perron, 1991.

Opsomer and Halleux 1985: Carmélia Opsomer and Robert Halleux, "La Lettre d'Hippocrate à Mécène et la lettre d'Hippocrate à Antiochus," in Innocenzo Mazzini and Franca Fusco (eds), *I testi di medicina latini antichi: problemi filologici e storici. Atti del I° Convegno internazionle Macerata – S. Severino M., 26–28 aprile 1984* (Università di Macerata pubblicazioni della Facolta di lettere e filosofia 28). Roma: Giorgio Bretschneider, 1985, pp. 341–364.

Oxford English Dictionary: *Oxford English Dictionary* Online. Oxford: Oxford University Press, 2023. https://www.oed.com/?tl=true.

Pansier 1907: Pierre Pansier, "Étude sur un manuscrit médical du XIe siècle," *Mémoires de l'Académie de Vaucluse* 2nd série 7 (1907), pp. 115–122.

Pansier 1924–1932: Pierre Pansier, *Histoire de la langue provençale à Avignon du XIe au XIXe siècle*, 5 volumes. Avignon: Aubanel, 1924–1932.

Paterson 1987: Linda M. Paterson, "La médecine en Occitanie avant 1250," in Peter T. Ricketts (ed.), *Actes du premier Congrès de l'Association internationale d'études occitanes.* London: AIEO and Westfield College, 1987, pp. 383–99.

Paterson 1988: Linda M. Paterson, "Military Surgery: Knights, Sergeants, and Raimon of Avignon's Version of the *Chirurgia* of Roger of Salerno (1180–1209)," in Christopher Harper-Bill and Ruth Harvey (eds), *The Ideals and Practice of Medieval Knighthood II: Papers from the Third Strawberry Hill Conference, 1986*. Woodbridge: Boydell & Brewer, 1988, pp. 117–146.

Pfeffer 1997: Wendy Pfeffer, *Proverbs in Medieval Occitan Literature*. Gainesville: University Press of Florida, 1997.

Pfeffer 2005: Wendy Pfeffer, "Lifting a Glass in Medieval Occitania," in Keith Busby, Bernard Guidot, and Logan E. Whalen (eds), *"De sens rassis": Essays in Honor of Rupert T. Pickens*. Amsterdam: Rodopi, 2005, pp. 529–542.

Pfeffer 2006: Wendy Pfeffer, "Christmas Gifts in Medieval Occitania: Matfre Ermengaud's Letter to His Sister," in Keith Busby and Christopher Kleinhenz (eds), *Courtly Arts and the Art of Courtliness: Selected Papers from the Eleventh Triennial Congress of the International Courtly Literature Society, University of Wisconsin-Madison, 29 July–4 August 2004*. Woodbridge: D. S. Brewer, 2006, pp. 517–525.

Pfeffer 2015: Wendy Pfeffer, "'Une noix, qu'y a-t-il à l'intérieur d'une noix?'" *Revue des langues romanes* 119 (2015), pp. 423–432.

Pfeffer 2016: Wendy Pfeffer, *Le Festin du troubadour: Nourriture, société et littérature en Occitanie, 1100–1500*. Cahors: La Louve, 2016.

Pfeffer 2019: Wendy Pfeffer, "Medieval Regionalism: Ingredients in the *Modus*," paper presented at the 5e Convention internationale d'histoire et des cultures de l'alimentation, Tours, France (June 6, 2019).

Pfeffer 2022: Wendy Pfeffer, "Expanding the +ATUM in Medieval Occitan: A Subcategory of Suffixed Words Relating to Foods," *Cultura Neolatina*, 82 (2022): pp. 13–24.

Pfeffer and Burrell 2022: Wendy Pfeffer (ed.) and Margaret Burrell (transl.), *Blandin de Cornoalha: A Comic Occitan Romance* (TEAMS, Varia Series). Kalamazoo, Michigan: Medieval Institute Publications, 2022.

Pfister 1960: Max Pfister, *Die Entwicklung der inlautenden Konsonnantengruppe -PS- in den romanischen Sprachen mit besonderer Berücksichtigung des Altprovenzalischen* (Romanica Helvetica 69). Bern: A. Francke, 1960.

Piñer 2022: Hélène Jawhara Piñer, *Jews, Food, and Spain: The Oldest Medieval Spanish Cookbook and the Sephardic Culinary Heritage*. Boston: Academic Studies Press, 2022.

Pucci Donati 2007: Francesca Pucci Donati, *Dieta, salute, calendari, dal regime stagionale antico ai regimina mensium medievali: origine di un genere nella letteratura medica occidentale* (Testi, studi, strumenti 22). Spoleto: Fondazione Centro italiano di studi sull'alto medioevo, 2007.

Raynouard 1836–1845: François-Juste-Marie Raynouard, *Lexique roman ou Dictionnaire de la langue des troubadours comparée avec les autres langues de l'Europe*, 6 volumes. Paris: Sylvestre, 1836–1845.

Redon et al. 1998: Odile Redon, Françoise Sabban, and Silvano Serventi. *The Medieval Kitchen: Recipes from France and Italy*, Edward Schneider (trans). Chicago: University of Chicago Press, 1998. Translation of *La Gastronomie au Moyen Âge: 150 recettes de France et d'Italie*. Paris: Stock, 1991.

Rei-Bèthvéder 2004: Nicolau Rei-Bèthvéder, *Dictionnaire français-occitan (gascon toulousain): Dictionnaire de la langue parlée en Nord Comminges, Fezensaguet, Lomagne, Muretain, Savès & pays toulousain: Gers, Haute-Garonne, Tarn-et-Garonne*. [Puylaurens]: Institut d'estudis occitans, 2004.

Réseau des bibliothèques universitaires de Montpellier: Réseau des bibliothèques universitaires de Montpellier, online catalogue. https://ged.biu-montpellier.fr/florabium/jsp/nodoc.jsp?NODOC=2015_DOC_MONT_MBUM_39.

Riché and Alexandre-Bidon 1994: Pierre Riché and Danièle Alexandre-Bidon, *L'enfance au Moyen Âge*. Paris: Seuil and Bibliothèque nationale de France, 1994.

Ricketts 1989: Peter T. Ricketts, "Plantes et recettes médicales dans le *Breviari d'amor* de Matfre Ermengaud de Béziers," in *Per Robert Lafont: Estudis ofèrts a Robert Lafont per sos collègas e amics*. Montpellier: Montpelhièr Universitat Paul-Valéry, Centre d'estudis occitans, 1990, pp. 255–266.

Riha 1987: Ortrun Riha, "Frühmittelalterliche Monatsdiätetik: Anmerkungen zu einem komplexen Thema," *Würzburger medizinhistorische Mitteilungen* 5 (1987), pp. 371–379.

Riha 1994: Ortrun Riha, "Die diätetischen Vorschriften der mittelalterlichen Monatsregeln," in Josef Domes, Bernhard D. Haage, Volker Zimmermann, and Werner E. Gerabek (eds), *Licht der Natur: Medizin in Fachliteratur und Dichtung, Festschrift für Gundolf Keil zum 60. Geburtstag* (Göppinger Arbeiten zur Germanistik 585). Göppingen: Deutscher Wissenschafters-Verlag, 1994, pp. 341–364.

Rinoldi 2021: Paolo Rinoldi, "Chirurgia volgare a Montpellier: il trattato occitano di Stefano Aldebaldi," *Carte Romanze* 9 (2021), pp. 177–198.

Rodet-Belardi and Forest 2009: Isabelle Rodet-Belardi and Vianney Forest, "Alimentation carnée du XIVe au XVIe siècle dans le sud de la France, d'après les sources archéozoologiques: la part des mammifères sauvages et des oiseaux," in Fabienne Ravoire and Anne Dietrich (eds), *La cuisine et la table dans la France de la fin du Moyen Âge. Colloque organisé par l'UMR 5594 (Dijon), l'INRPA, et le Centre d'étude et de recherche du patrimoine de Sens (Sens, 8–10 janvier 2004)*. Caen: Publications du CRAHM, 2009, pp. 125–146.

Ronjat 1930–1941: Jules Ronjat, *Grammaire istorique des parlers provençaux modernes*, 4 volumes. Montpellier: Société des langues romanes, 1930–1941.

Rothwell 1983: William Rothwell, "Medical and Botanical Vocabulary in Old Provençal: Some Considerations," in Manuel Jorba (ed.), *Estudis de llengua i literatura catalanes oferts a R. Aramon i Serra en el seu setantè aniversari*, volume 3 (Estudis universitaris catalans 23). Barcelona: Curial, 1983, pp. 489–502.

Ruas 1992: Marie-Pierre Ruas, "Les plantes exploitées en France au Moyen Âge d'après les semences archéologiques," in *Plantes et cultures nouvelles en Europe occidentale au Moyen Âge et à l'époque moderne. Centre culturel de l'Abbaye de Flaran, douzièmes journées internationales d'histoire, 11, 12, 13 septembre 1990*. Auch: Comité départemental du tourisme du Gers, 1992, pp. 9–35.

Ruas et al. 2005–2006: Marie-Pierre Ruas, Laurent Bouby, Perrine Mane, Carole Puig, and Bénédicte Pradat, "Les fruits de l'alimentation médiévale en France du sud, entre marchés, recettes et dépotoirs," *Archéologie du Midi médiéval*, 23–24 (2005–2006), pp. 195–206.

Ruggieri 1930: Yole Ruggieri, "La 'Dietetica' provenzale del ms. Vat. Barb. 311," *Atti della Reale Accademia della Scienze di Torino* 65 (1930), pp. 203–219.

Sabbah et al. 1987: Guy Sabbah, Pierre-Paul Corsetti, and Klaus-Dietrich Fischer (eds), *Bibliographie des textes médicaux latins: Antiquité et haut Moyen Âge* (BTML). Preface by Mirko D. Grmek (Centre Jean-Palerne, Mémoires 6). Saint-Étienne: Publications de l'Université de Saint-Étienne, 1987.

Santich 2018: Barbara Santich, *The Original Mediterranean Cuisine: Medieval Recipes for Today*, 2nd ed. Sheffield and Bristol, Connecticut: Equinox Publishing, 2018.

Segre 1968: Cesare Segre, "B. Le forme e le tradizione didattiche," in Jürgen Beyer (ed.), *La littérature didactique, allégorique et satirique*, volume 1, in Hans Robert Jauss (dir.), *Grundriß der romanischen Literaturen des Mittelalters*, volume 6. Heidelberg: Carl Winter Universitätsverlag, 1968, pp. 58–145.

Smith 1976: Nathaniel B. Smith, *Figures of Repetition in the Old Provençal Lyric: A Study in the Style of the Troubadours* (North Carolina Studies in the Romance Languages and Literatures 176). Chapel Hill: University of North Carolina, Department of Romance Languages, 1976.

Stouff 1969: Louis Stouff, "La viande. Ravitaillement et consommation à Carpentras au XVème siècle," *Annales: économies, sociétés, civilisations* 24 (1969): 1431–1448.

Stouff 1970: Louis Stouff, *Ravitaillement et alimentation en Provence aux XIVe et XVe siècles.* Paris and The Hague: Mouton, 1970.

Stouff 1996: Louis Stouff, *La table provençal: boire et manger en Provence à la fin du Moyen Âge*. Avignon: A. Barthélemy, 1996.

Suchier 1883: Hermann Suchier (ed.), *Denkmäler provenzalischer Literatur und Sprache zum ersten Male herausgageben.* Halle: Max Niemeyer, 1883.

Sudhoff 1918a: Karl Sudhoff, "Aus einer Sammlung chirurgischer Rezepte aus dem 13. Jahrhundert auf der Vallicellana in Ro," *Studien zur Geschichte der Medizin* 11 (1918), pp. 395–390.

Sudhoff 1918b: Karl Sudhoff, "Domini et magistri Willehelmi de Congenis, Burgensis zu Montpellier Scriptum cirurgiae, sovie dazugehörige 'Notulae cirurgiae' eines Schulers," *Studien zur Geschichte der Medizin* 11 (1918), pp. 297–385.

Swisscollections: Swissollections. https://aleph.unibas.ch/F/?local_base=DSV05&con_lng=GER&func=find-b&find_code=SYS&request=000123454.

Talbot 1972: Charles H. Talbot, "Medico-Historical Introduction," in Charles H. Talbot and Franz Unterkircher, *Medicina antiqua: Libri quattuor medicinae, Codex Vindobonensis 95 der österreichischen Bibliothek* (Codices selecti phototypice impressi 27 and 27*), volume 1. Graz: Akademische Druck- und Verlagsanstalt, 1972, pp. 11–35.

Talfani 2022: Camilla Talfani, "La *scripta* du Languedoc occidental et la *scripta* de la Provence au XIVe siècle," in Fabio Barberini and Camilla Talfani (eds), *Trans-mission, création et hybridation dans le domaine d'oc: nouvelles perspectives de la recherche en domaine occitan* (Publications de l'Association internationale d'études occitanes 14). Turnhout: Brepols, 2022, pp. 149–165.

Taylor 2015: Robert A. Taylor, *Bibliographical Guide to the Study of the Troubadours and Old Occitan Literature*. Kalamazoo, Michigan: Medieval Institute Publications, 2015.

Temkin 1973: Owsei Temkin, *Galenism: Rise and Decline of a Medical Philosophy* (Cornell Publications in the History of Science). Ithaca, NY, and London: Cornell University Press, 1973.

Thorndike 1923: Lynn Thorndike, *A History of Magic and Experimental Science*, volume 2. New York, NY: Columbia University Press and Macmillan, 1923.

Thorndike and Kibre 1963: Lynn Thorndike and Pearl Kibre, *A Catalogue of Mediaeval Scientific Writings in Latin*, revised and augmented edition. Cambridge, Massachusetts: Medieval Academy of America, 1963.

Tovar 1974: Claude de Tovar, "Contamination, interférences et tentatives de systématisation dans la tradition manuscrite des réceptaires médicaux français: le réceptaire de Jean Sauvage," *Revue d'histoire des textes* 3 (1974), pp. 115–191.

Van d'Elden 1991: Stephanie Cain Van d'Elden, "The Salerno Effect: The Image of Salerno in Courtly Literature," in Giovanna Angeli and Luciano Formisano (eds), *L'imaginaire courtois et son double, Actes du XIe Congrès international de la Société internationale de littérature courtoise (ICLS), Fisciano (Salerno), 24–28 juillet 1989* (Pubblicazioni dell'Università degli Studi di Salerno, Sezione Atti, Convegni, Miscellanee 35). Napoli: Edizioni Scientifiche Italiane, 1991, pp. 503–515.

Van der Horst 1981: Cor[nelius] Van der Horst, "Examen dialectologique du manuscrit P_2 des *Évangiles de l'enfance* en Occitan," *Zeitschrift für romanische Philologie* 97 (1981), pp. 329–351.

Van Winter 2010: Johanna Maria Van Winter, "*Regimina sanitatis* by Month and by Season," in Patricia Lysaght (ed.), *Food and Meals at Cultural Crossroads. Proceedings of the 17th Conference of the International Commission for Ethnological Food Research, Oslo, Norway, September 15–19, 2008*. Oslo: Novus Press, 2010, pp. 98–109.

Vayssier 1878: Aimé Vayssier, *Dictionnaire patois-français du département de l'Aveyron*. Rodez: Veuve E. Carrère, 1879.

Vielliard 2009: Françoise Vielliard, "Les traductions des classiques dans la littérature médiévale occitane," in *L'Occitan, une langue du travail et de la vie quotidienne du XIIe au XXIe siècle. Les traductions et les termes techniques en langue d'oc. Actes du colloque organisé à Limoges les 23 et 24 mai 2008 par le Centre Trobar et l'EA 4116*. Ussel: Musée du Pays d'Ussel and Centre trobar, 2009, pp. 223–237.

Wackernagel 1845: Wilhelm Wackernagel, "Provenzalische Diätetik," *Zeitschrift für deutsches Alterthum* 5 (1845), pp. 16–17.

Wallis 1995: Faith Wallis, "Medicine in Medieval Calendar Manuscripts," in Margaret R. Schleissner (ed.), *Manuscript Sources of Medieval Medicine: A Book of Essays* (Medieval Casebooks Series 8). New York: Routledge, 1995, pp. 105–143.

Wallis 2010: Faith Wallis, *Medieval Medicine: A Reader* (Readings in Medieval Civilizations and Cultures 15). Toronto: University of Toronto Press, 2010.

Wartburg 1922–2002: Walter von Wartburg (ed.), *Französisches etymologisches Wörterbuch: eine Darstellung des galloromanischen Sprachschatzes*, 25 vols. Tübingen, then Basel: J. C. B. Mohr, then Helbing und Lichtenhahn, then Zbinden, 1922–2002.

Weisse 1883: Richard Weisse, *Die Sprachformen Matfre Ermengau's*. Inaugural-Dissertation … der Vereinigten Friedrichs-Universität Halle-Wittenberg. Halle an der Saale: E. Karras 1883.

Williams 2003a: Steven J. Williams, *The Secret of Secrets: The Scholarly Career of a Pseudo-Aristotelian Text in the Latin Middle Ages*. Ann Arbor: University of Michigan Press, 2003.

Williams 2003b: Steven J. Williams, "The Vernacular Tradition of the Pseudo-Aristotelian *Secret of Secrets* in the Middle Ages: Translations, Manuscripts, Readers," in Nadia Bray and Loris Sturlese (eds), *Filosofia in volgare nel Medioevo. Atti del Convegno della Società italiana per lo studio del pensiero medievale (S. I. S. P. M.), Lecce, 27–29 settembre 2002* (Textes et études du Moyen Âge 21). Louvain-la-Neuve: Féderation Internationale des Instituts d'Études Médiévales, 2003, pp. 451–482.

Williams 2021: Steven J. Williams, "The Pseudo-Aristotelian *Secret of Secrets*: Encyclopedia or Book of Knowledge?" in Pavlina Cermanová and Václav Žůrek (eds), *Books of Knowledge in Late Medieval Europe: Circulation and Reception of Popular Texts* (Utrecht Studies in Medieval Literacy 52). Turnhout: Brepols, 2021, pp. 11–34.

World Health Organization 2010: World Health Organization, *WHO Guidelines on Drawing Blood: Best Practices in Phlebotomy*. Geneva: World Health Organization, 2010. https://www.who.int/publications/i/item/9789241599221.

Zamuner [Candiani] 2003: Ilaria Zamuner [Candiani], "Per l'edizione critica dei volgarizzamenti provenzali dell' 'Epistola ad Alexandrum de dieta servanda,'" in Rossana Castano, Saverio Guida, and Fortunata Latella (eds), *Scène, évolution, sort de la langue et de la littérature d'oc. Actes du Septième Congrès international de l'Association internationale d'études occitanes, Reggio Calabria–Messina, 7–13 juillet 2002*, volume 1. Rome: Viella, 2003, pp. 739–759.

Zamuner [Candiani] 2004: Ilaria Zamuner [Candiani], "Il ms. Barb. Lat. 311 e la trasmissione dei *regimina sanitatis* (XIII–XV sec.)," *Cultura Neolatina* 64 (2004), pp. 207–250.

Zamuner [Candiani] 2005: Ilaria Zamuner [Candiani], "La tradizione romanza del *Secretum secretorum* pseudo-aristotelico. Regesto delle versioni e dei manoscritti," *Studi Medievali*, 3rd series, 46 (2005), pp. 31–116.

Zamuner [Candiani] 2006: Ilaria Zamuner [Candiani], "Les versions françaises de l'*Epistola ad Alexandrum de dieta servanda*: mise au point," in Claudio Galderisi and Cinzia Pignatelli (eds), *La traduction vers le moyen français. Actes du IIe colloque de l'AIEMF [Association internationale pour l'étude du moyen français], Poitiers, 27–29 avril, 2006*. Turnhout: Brepols, 2007, pp. 165–187.

Zamuner [Candiani] 2010: Ilaria Zamuner [Candiani], "Un nuovo testimone della Chirurgia di Ruggero Frugardo in lingua occitanica (Siviglia, Biblioteca Columbina, ms. 5-5-20)," in Anna Alberni, Lola Badia, and Lluís Cabré (eds), *Translatar i transferir: La transmissió dels textos i el saber (1200–1500). Actes del Premier Col·loqui internacional del grup NARPAN, Barcelona, 22–23 novembre 2007.* Santa Coloma de Queralt: Obrador Edèndum, 2010, pp. 191–240.
Zamuner [Candiani] 2021: Ilaria Zamuner [Candiani], "Una versione veneziana dell'*Epistola ad Alexandrum de dieta servanda*," in Anna Alberni, Lluís Cifuentes, Joan Santanach, and Albert Soler (eds), *"Qui fruit ne sap collir". Homenatge a Lola Badia*, volume 2. Barcelona: Edicions de la Universitat de Barcelona, 2021, pp. 347–364.
Zamuner [Candiani] forthcoming: Ilaria Zamuner [Candiani], "I ricettari medici in lingua occitanica: stato dell'arte e analisi di un caso particolare (con esempi di lessicologia comparata)," in Catherine Menichetti (ed.), *La littérature occitane médiévale dans sa tradition manuscrite. Témoins, traditions, corpora. Atti del Convegno (Losanna, 30 novembre–3 dicembre 2021).* Rome: Viella, forthcoming.
Zufferey 1987: François Zufferey, *Recherches linguistiques sur les chansonniers provençaux* (Université de Lausanne, Faculté des lettres). Genève: Droz, 1987.

Websites

ARLIMA, Archives de littérature du Moyen Âge. https://www.arlima.net/index.html.
Calames, Catalogue en ligne des archives et des manuscrits de l'enseignement supérieur. http://www.calames.abes.fr.
CoReMA, Cooking Recipes of the Middle Ages. https://gams.uni-graz.at/context:corema.
Dictionary of Old Occitan Medico-Botanical Terminology (DiTMAO). https://www.uni-goettingen.de/en/487498.html.
Dictionnaire de l'occitan médiéval online. www.dom-en-ligne.de/.
Dictionnaires occitans. https://dicodoc.eu/fr/dictionnaires.
Gallica, Bibliothèque nationale de France. https://gallica.bnf.fr/accueil/en/content/accueil-en?mode=desktop.
Gran Diccionari de la llengua catalana. https://www.diccionari.cat/gran-diccionari-de-la-llengua-catalana#.
Institut d'Estudis Catalans, *Diccionari català-valencià-balear.* https://dcvb.iec.cat/.
Jonas database. https://jonas.irht.cnrs.fr.
National Institutes of Health. https://www.ncbi.nlm.nih.gov.
RBME Digital, Patrimonio Nacional, online publication works of San Lorenzo de El Escorial, Real Biblioteca del Monasterio. https://rbme.patrimonionacional.es/s/rbme/page/introducci-n.
Swissollections. https://aleph.unibas.ch/F/?local_base=DSV05&con_lng=GER&func=find-b&find_code=SYS&request=000123454.
Trésor manuscrit de l'ancien occitan (TMAO). http://tmao.aieo.org/. Access to the Trésor is restricted to members of the Association internationale d'études occitanes.
Wikipedia. https://www.wikipedia.org/.
World Health Organization. https://www.who.int/publications/i/item/9789241599221.

Indexes

References throughout the indexes are to pages, footnotes, or to the Occitan text. Notes are identified with a page locator, followed by the letter “n” and the note number (e.g. 60n346 refers to note 346 on page 60). References to a line or lines in the Occitan text or its translation are identified by the word “text” followed by the line number in the Occitan text.

Toponyms and Personal Names

https://doi.org/10.1515/9783111268125-004

Subject Index